BARRON'S

NEW JERSEY

GRADE 5

MATH TEST

Stephenie Tidwell, M.A.

MW00563915

About the Author

Stephenie Tidwell is a District Supervisor of K–5 Mathematics in a public school district in New Jersey, where she provides ongoing professional development opportunities that address teacher instructional practice. Stephenie joined the field of education after a 16-year tenure in the Banking and Finance Industry. She holds two master's degrees from Montclair State University; one in Educational Leadership and the other in Social Science/Economics. She credits her undergraduate economics professors at North Carolina Agricultural and Technical State University for recognizing her passion for statistical analysis and problem solving. Stephenie resides in Union where she enjoys golfing, ballroom dancing, and spending time with her family.

Acknowledgments

I would like to thank my parents and constant cheerleaders, Ponce and Marva Tidwell, for their continuous nonstop support. I must also thank those who currently challenge and push me to new heights—Dr. Eric Milou, Dr. Linda Gross, Rev. Dr. Mattie Smith, Mr. John Scozzaro, and Mrs. Barbara Taylor. Thanks also go to my family and friends who believe in and support my efforts—Ponce (Jr.), G.H. Brooks, Donna, Ernie, Felicia, Whitney, Dawn, Andrea, Vanessa, Kim, Linda, Paula, Hanifa, Karen, Bert, Aisha, and all others I could not name due to limited space.

© Copyright 2015 by Barron's Educational Series, Inc.

All rights reserved.
No part of this publication may be reproduced or distributed in any form or by any means without the written permission of the copyright owner.

All inquiries should be addressed to:
Barron's Educational Series, Inc.
250 Wireless Boulevard
Hauppauge, NY 11788
www.barronseduc.com

ISBN: 978-1-4380-0722-9
Library of Congress Control Number: 2015935025

Date of Manufacture: October 2015
Manufactured by: B11R11

Printed in the United States of America
9 8 7 6 5 4 3 2 1

CONTENTS

INTRODUCTION

Imagine that it's your first day of fifth-grade math. You are excited and nervous at the same time. The teacher seems nice. She asks the class, "What languages are you fluent in, other than English? Talk a moment in your groups." She walks from group to group. Students are saying proudly, "Spanish, French, Arabic, and Hindi." She calls time and expresses that she has a wonderfully diverse classroom; however, not one student mentioned that he or she was fluent in the language of mathematics. Just imagine the confused looks on your faces? Your math teacher begins to explain how math assists us in expressing ourselves numerically. She then describes how the Common Core State Standards are designed to assist you in developing fluency in the standards across grade levels in an effort to prepare you to become college and career ready. She states, "To become fluent in the language of mathematics you must practice speaking the language." This New Jersey Grade 5 Math Test Book has been created to assist you in developing an understanding of mathematics by practicing the language of the math standards and assessing your fluency in the standards as you prepare for standardized testing.

—Stephenie Tidwell

As a former bank administrator turned mathematics teacher, I recognized the importance of helping my students to understand that mathematics was all around them. As a banker, I assisted many adults in solving problems that involved basic math skills. For some reason, they could not connect the math from their younger years to the real-world tasks required of their current job. These situations only helped to solidify my prior thought. I was supposed to be and wanted to be a math teacher. Not only could I bring real-world math problems into the math classroom to help answer the continuous question of "Why do we have to learn this math stuff anyway?" I could also assist students in understanding the language of mathematics and in making the connection between math and their everyday lives.

The purpose of this book is to assist parents, students, and teachers in understanding that mathematics is truly a language of its own that can be spoken by anyone given it is studied, practiced, and understood. The language of mathematics is not based upon some mutant gene that is inherited across generations; as I have often heard repeated from the mouths of my students as they tend to try and justify their mathematical existence or lack thereof by stating "My parents were never good at math, so I'm not good at math" or "My mom/dad was good in math so I must have gotten it from her/him."

The Common Core State Standards are designed to provide students with a solid mathematical foundation and understanding of mathematical concepts as they progress across grade levels. This book has been designed specifically to assist fifth-grade students in understanding what the Common Core Standards mean for each domain while connecting the mathematical concepts to their lives.

This book allows students to practice the language of mathematics for understanding. Proper implementation of the standards and use of this book will assist students in developing an understanding of mathematics as they work towards becoming fluent in the language of mathematics.

The New Jersey Math Assessment PARCC Grade 5

CHAPTER 1

Book Overview

The Barron's New Jersey PARCC Grade 5 Math Test workbook is a test-prep tool to assist students in preparing for the PARCC Assessment. The book includes an overview of the Common Core Standards and their connection to the PARCC Assessment. Students will have the opportunity to take a diagnostic assessment to identify the standards in which they have obtained mastery or require additional review work and support. The book has five review sections organized by the domain, with specific focus on the major clusters for the grade level (see Table 1 on pages 5 and 6). Each review section has student-friendly explanations of the standards[1] and review lessons with examples followed by PARCC Type I—Procedure, Type II—Reasoning, and Type III—Application-level questions and the corresponding answers. The final two chapters of the workbook contain two PARCC-like practice assessments with questions that model the *PARCC High Level Blueprints for Mathematics.* Additionally, Appendix A contains a list of math websites that can be utilized for additional review and practice problems.

Common Core Overview

Imagine that you moved from Connecticut to New Jersey. You are excited about going to a new school and all the new things you will learn in math class. It's your first day in your new school, you are listening to your teacher as she tells the class everything they can expect to learn this year, and you realize that you learned most of what she's saying last year. Additionally, imagine your teacher thinking, "Okay. Hmmm? I wonder what topics his/her teacher covered last year and to what level of understanding?"

Well, that was math education prior to the 2009 state-led efforts to develop consistent, real-world learning goals called the Common Core State Standards. This initiative was spear-headed by state leaders, including governors and state commissioners of education from 48 states, two territories, and the District of Columbia through their membership in the National Governors Association (NGA) and the Council of Chief State School Officers (CCSSO). The primary goal of educators was to ensure that all students, regardless of where they live, are introduced to the same learning goals and objectives for their particular grade level so that graduation from high school now means students

[1] Additional elaboration of certain standards will be available when necessary.

are prepared for college, career, and life. How's that possible, you ask? Through a lot of hard work and collaborative efforts, a set of consistent standards were created based on:

- Identifying the best state standards already in existence
- The experience of teachers, content experts, and state departments of education
- Feedback from the public

(Source: *http://www.corestandards.org/about-the-standards/development-process/*)

Fast forward, it's now June 2010 and the New Jersey State Board of Education (NJBOE) along with the New Jersey Department of Education (NJDOE) adopts the Common Core State Standards for mathematics to replace the previous math standards. Why? Primarily because the Common Core State Standards allow every parent and teacher to support student learning. How? Through standards that define what students should know and be able to do at each grade level, establishing a clear and consistent framework to prepare children for college and the workforce. (Source: *http://www.state.nj.us/education/sca/*)

Common Core Introduction

In Grade 5, your instructional time should focus on three critical areas: (1) developing fluency with addition and subtraction of fractions and developing an understanding of the multiplication of fractions and the division of fractions in limited cases (unit fractions divided by whole numbers and whole numbers divided by unit fractions); (2) extending your fourth-grade knowledge of division to include division to two-digit divisors, incorporating decimal fractions into the place value system, and developing your understanding of operations with decimals to hundredths, along with developing your fluency with whole number and decimal operations; and (3) developing an understanding of volume.

1. Students apply their understanding of fractions and fraction models to represent the addition and subtraction of fractions with unlike denominators as equivalent calculations with like denominators. They develop fluency in calculating sums and differences of fractions and make reasonable estimates of them. Students also use the meaning of fractions, the meaning of multiplication and division, and the relationship between multiplication and division to understand and explain why the procedures for multiplying and dividing fractions make sense. (*Note*: This is limited to the case of dividing unit fractions by whole numbers and whole numbers by unit fractions.)
2. Students develop an understanding of why division procedures work based on the meaning of base-ten numerals and the properties of operations. They finalize fluency with multi-digit addition, subtraction, multiplication, and division. They apply their understandings of models for decimals, decimal

notation, and properties of operations to add and subtract decimals to hundredths. They develop fluency in these computations and make reasonable estimates of their results. Students use the relationship between decimals and fractions, as well as the relationship between finite decimals and whole numbers (i.e., a finite decimal multiplied by an appropriate power of 10 is a whole number), to understand and explain why the procedures for multiplying and dividing finite decimals make sense. They compute products and quotients of decimals to hundredths efficiently and accurately.

3. Students recognize volume as an attribute of three-dimensional space. They understand that volume can be measured by finding the total number of same-size units of volume required to fill the space without gaps or overlaps. They understand that a 1-unit by 1-unit by 1-unit cube is the standard unit for measuring volume. They select appropriate units, strategies, and tools for solving problems that involve estimating and measuring volume. They decompose three-dimensional shapes and find volumes of right rectangular prisms by viewing them as decomposed into layers of arrays of cubes. They measure necessary attributes of shapes in order to determine volumes to solve real-world and mathematical problems.
(Source: *http://www.corestandards.org/Math/Content/5/introduction/*)

Mathematical Content

The fifth-grade Common Core Standards consists of five domains. The domains can be thought of as the main topic or theme of the unit. Under the main topic, you have clusters of standards. The clusters can be thought of as the big ideas for each domain. These clusters or big ideas are groupings of individual standards that address what you need to know and be able to do by the end of the grade level (see Table 1).

Table 1. Grade 5 Common Core State Standards Overview

Domain: Operations and Algebraic Thinking	
	• Write and interpret numerical expressions. • Analyze patterns and relationships.

<table>
<tr><th colspan="2">Domain: Number and Operations in Base 10</th></tr>
<tr><td></td><td>• Understand the place value system.
• Perform operations with multi-digit whole numbers and with decimals to hundredths.</td></tr>
<tr><th colspan="2">Domain: Number and Operations—Fractions</th></tr>
<tr><td></td><td>• Use equivalent fractions as a strategy to add and subtract fractions.
• Apply and extend previous understandings of multiplication and division to multiply and divide fractions.</td></tr>
<tr><th colspan="2">Domain: Measurement and Data</th></tr>
<tr><td></td><td>• Convert like measurement units within a given measurement system.
• Represent and interpret data.
• Geometric measurement: understand concepts of volume and relate volume to multiplication and to addition.</td></tr>
<tr><th colspan="2">Domain: Geometry</th></tr>
<tr><td>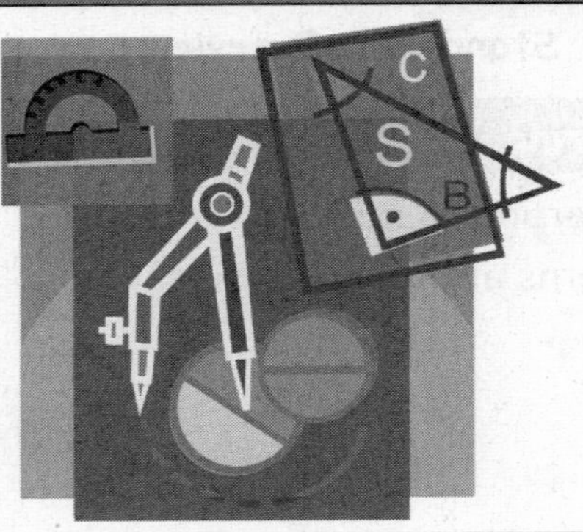</td><td>• Graph points on the coordinate plane to solve real-world and mathematical problems.
• Classify two-dimensional figures into categories based on their properties.</td></tr>
</table>

Standards for Mathematical Practice

It's one thing to be able to do the mathematics required to provide the answer to a math problem. However, it's another thing to develop your thinking about the mathematics. The Standards for Mathematical Practice have been established to ensure that you are a well-rounded mathematical thinker and communicator. As a mathematical thinker, you make sense of the problem and work at it until it is solved; you are able to organize your ideas, explain your thinking, or identify incorrect thinking; you are able to draw pictures or use models to represent the problem; you are able to determine the appropriate tools (a calculator, ruler, protractor) to use; and you can identify patterns and then use the patterns to assist you in solving the problem. Table 2 provides the complete list of the Standards for Mathematical Practice.

Table 2. Standards for Mathematical Practice

MP 1	Make sense of problems and persevere in solving them	You determine what the problem is asking and look for efficient ways to represent and solve it. You do not give up until you have found the solution. You may check your thinking by asking yourself, "What is the most efficient way to solve the problem?", "Does this make sense?", and "Can I solve the problem in a different way?"
MP 2	Reason abstractly and quantitatively	You recognize that a number represents a specific quantity. You are able to connect quantities to written symbols and create an equation to represent the problem at hand. You make mathematical assumptions and evaluate them to solve the problem.
MP 3	Construct viable arguments and critique the reasoning of others	You may use objects, pictures, or drawings to communicate or defend your mathematical reasoning. You can verbally and in writing explain your mathematical thinking, how you arrived at an answer, or why your answer is true. You are able to evaluate and respond to others' thinking.
MP 4	Model with mathematics	You are able to represent problem situations in multiple ways including numbers, words (mathematical language), pictures, objects, charts, lists, graphs, equations, etc. You should be able to determine which models are most useful and efficient to solve problems.

Table 2. *Continued*

MP 5	Use appropriate tools strategically	You consider the available tools (including estimation) when solving a mathematical problem and decide when certain tools might be helpful to solve a specific problem. For example, you may use unit cubes to determine the volume of a rectangular prism and then use a ruler to measure the length, width, and height.
MP 6	Attend to precision	You use clear and precise mathematical language in your discussions with others and in your own reasoning. You use the appropriate math symbols when solving problems. You are careful about specifying the correct units of measure. For example, you use cubic units to answer a question about volume.
MP 7	Look for and make use of structure	You look closely to discover a pattern or structure. You examine numerical patterns and relate them to a rule or a graphical representation.
MP 8	Look for and express regularity in repeated reasoning	You use repeated reasoning to understand algorithms and make generalizations about patterns. For example, you explore operations with fractions with visual models and begin to formulate conclusions.

What Is PARCC?

The Partnership for Assessment of Readiness for College and Careers (PARCC) is a state consortium responsible for developing high-quality student assessments aligned to the new Common Core State Standards (CCSS). This assessment replaces the pencil-and-paper-based NJASK assessment with a new computer-based assessment. Unlike the NJASK, the PARCC assessment will require you to complete tasks designed to assess your use of the strategies and explanations essential to the CCSS. The PARCC assessment contains various types of questions such as multiple-choice questions requiring more than one selection; short-constructed response questions requiring the only answer; and extended-constructed response questions that require you to show your work, identify strategies used, and explain the process to arrive at an answer. Additionally, the assessment will require you to use the online manipulatives and tools to demonstrate your mathematical understanding of certain concepts.

Barron's has made every effort to ensure that the content of this book is accurate as of press time; however, the PARCC Assessments are constantly changing. Your teacher and parents can stay informed of the latest testing information by registering to receive PARCC updates at the following website: *http://www.parcconline.org/*. Regardless of the changes that may be announced after press time, this book will still provide a strong framework for fifth-grade students' preparation for the assessment.

What Will PARCC Math Assessment Look Like?

As of the publication date of this book the initial PARCC Math Assessment has been adapted and will now be given in a 30-day testing window. This new single testing window will extend from roughly the 75% to the 90% mark of the school year which should allow for schools to complete testing within one to two weeks during that window. The PARCC assessment new test design changes will not result in any loss of the performance tasks and will still contain three different types of questions:

Type I–Tasks assessing concepts, skills, and procedures

- These tasks are a balance of conceptual understanding, fluency, and application
- Can involve any or all of the mathematical practice standards
- Will be machine scorable with innovative, computer-based formats

Type II–Tasks assessing and expressing mathematical reasoning

- These tasks call for written arguments/justifications, critique of reasoning or precision in mathematical statements (MP 3, 6)
- Can involve other mathematical practice standards
- May include a mix of machine-scored and hand-scored responses

Type III–Tasks assessing modeling/applications

- These tasks call for modeling/application in a real-world context or scenario (MP 4)
- Can involve other mathematical practice standards
- May include a mix of machine-scored and hand-scored responses

If you had to look at the different question types as that of building a house, you could imagine that the structure of the house would look like this:

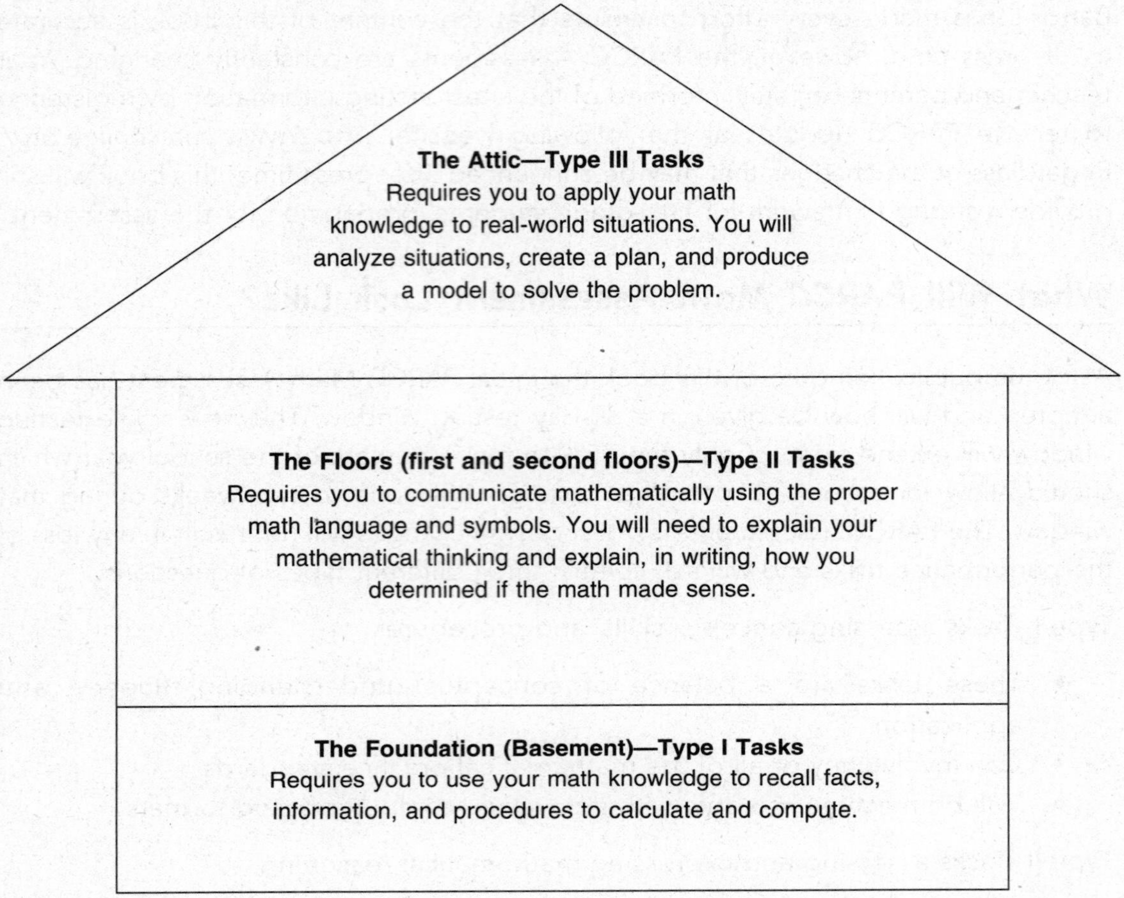

The Attic—Type III Tasks
Requires you to apply your math knowledge to real-world situations. You will analyze situations, create a plan, and produce a model to solve the problem.

The Floors (first and second floors)—Type II Tasks
Requires you to communicate mathematically using the proper math language and symbols. You will need to explain your mathematical thinking and explain, in writing, how you determined if the math made sense.

The Foundation (Basement)—Type I Tasks
Requires you to use your math knowledge to recall facts, information, and procedures to calculate and compute.

The redesigned PARCC Assessment will be administered in multiple sections called units. The proposed 2015–2016 math assessment will have three or four units with a test time of one hour per unit. As of press time, the final mathematics unit configuration is still being considered.

Sample PARCC Math Assessment Items

Unlike testing of old, the PARCC assessment is totally computerized. As such, assessment items will require you to have an understanding of using a mouse and keyboard. Assessment items may require you to input letters, numbers, and mathematical symbols like <, >, =. Additionally, questions will also require students to drag and drop information to an answer box, select two or more answers as a response, and/or complete two-part questions that may require some additional calculations on your part. The following sample items give you an idea of the type of problems that can appear on the test.

Type I Sample Question

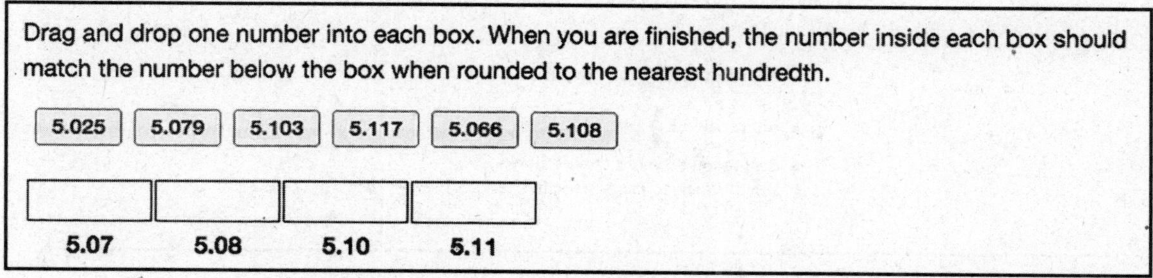

Drag and drop one number into each box. When you are finished, the number inside each box should match the number below the box when rounded to the nearest hundredth.

5.025 | 5.079 | 5.103 | 5.117 | 5.066 | 5.108

5.07 | 5.08 | 5.10 | 5.11

Type I Sample Question

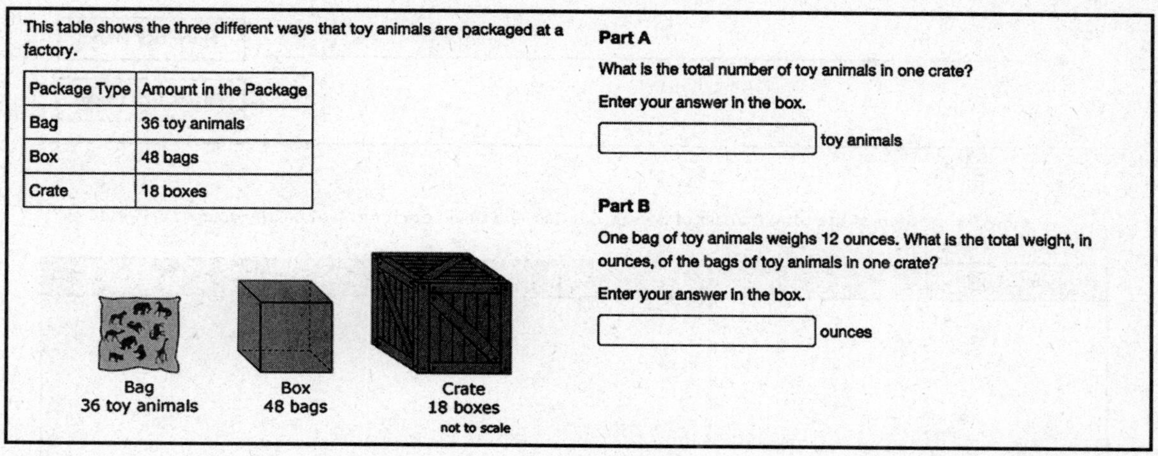

This table shows the three different ways that toy animals are packaged at a factory.

Package Type	Amount in the Package
Bag	36 toy animals
Box	48 bags
Crate	18 boxes

Bag
36 toy animals

Box
48 bags

Crate
18 boxes
not to scale

Part A

What is the total number of toy animals in one crate?

Enter your answer in the box.

______ toy animals

Part B

One bag of toy animals weighs 12 ounces. What is the total weight, in ounces, of the bags of toy animals in one crate?

Enter your answer in the box.

______ ounces

Type II Sample Question

Ryan added two fractions and got a sum of $\frac{3}{7}$.

$$\frac{1}{4}+\frac{2}{3}$$

Use your knowledge of fractions to explain why Ryan's answer is incorrect.

Type III Sample Question

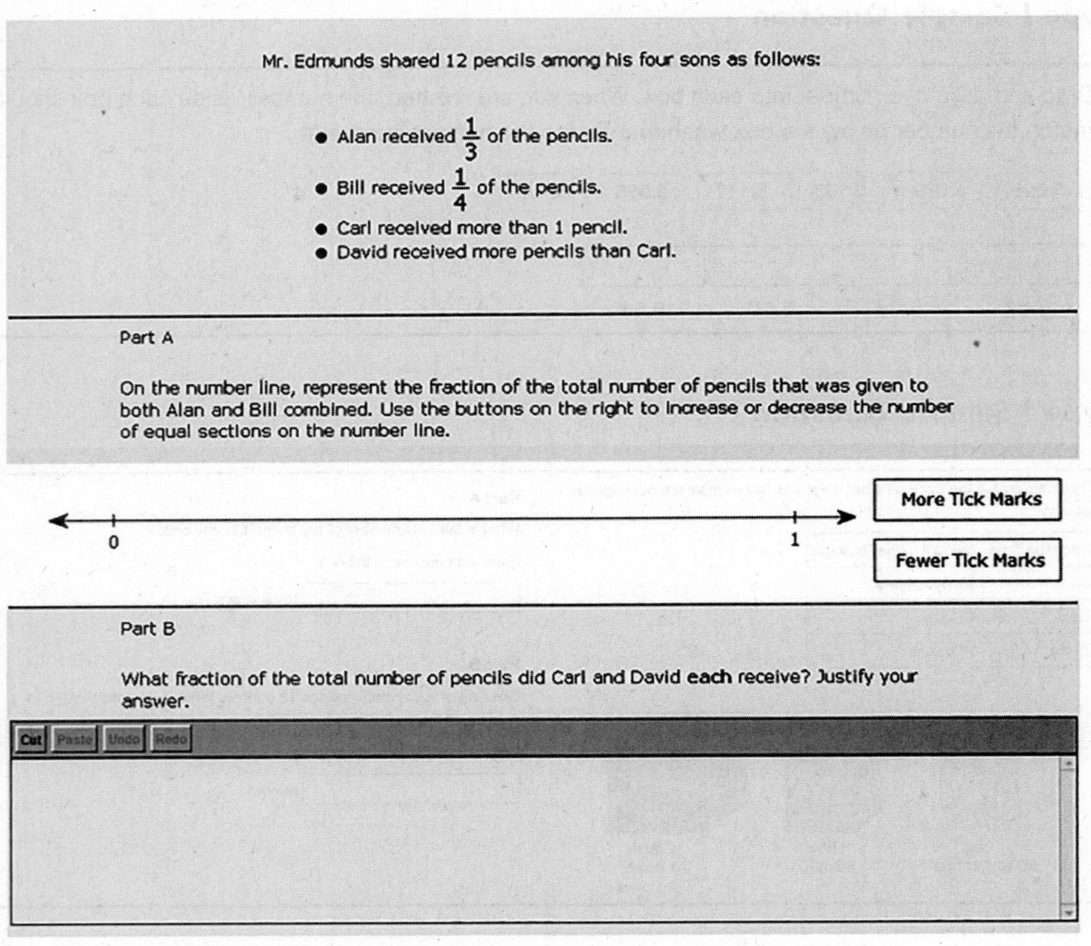

Diagnostic Assessment

CHAPTER 2

How to Use the Assessment

The diagnostic questions are similar, in format, to the test items as they will appear on the PARCC assessment. This diagnostic assessment is provided to assist you in identifying your strengths as weaknesses in the specific domains. You can use the results to improve your skills in the domains, as well as your speed and accuracy in test taking.

HELPFUL HINT:

Make sure to read the question carefully and to use the directions given in the question. Do your best to answer all the questions. However, if you do not know the answer to a question, skip it and go on to the next question. Before submitting your test, go back and answer all the skipped questions.

DIAGNOSTIC ASSESSMENT

Begin Here:

Your name:

Time started:

Time ended:

You may not use a calculator on this assessment.

1. Select the **two** answers that correctly identify the value of the underlined digits below.

1.$\underline{7}$8 and 1.8$\underline{7}$

□ A. The value of the 7 in the tenths place is 100 times as much as the value of the 7 in the hundredths place.

□ B. The value of the 7 in the tenths place is 10 times as much as the value of the 7 in the hundredths place.

□ C. The value of the 7 in the hundredths place is $\frac{1}{10}$ the value of the 7 in the tenths place.

□ D. The value of the 7 in the hundredths place is $\frac{1}{100}$ the value of the 7 in the tenths place.

2. The number 30,000 can also be expressed as:

○ A. $3 \times 10 \times 10 \times 10 = 3 \times 10^3$

○ B. $3 \times 10 \times 10 \times 10 \times 10 \times 10 = 3 \times 10^5$

○ C. $3 \times 10 \times 10 \times 10 \times 10 = 3 \times 10^4$

○ D. $3 \times 10 \times 10 = 3 \times 10^2$

3. Select the **two** ways to represent the number

721.369

□ A. Seven two hundred and one three hundred sixty nine hundredths

□ B. 700 + 20 + 1 + 0.3 + 0.06 + 0.009

□ C. $(72 \times 100) + (1 \times 10) + \left(3 \times \frac{1}{10}\right) + \left(6 \times \frac{1}{100}\right) + \left(9 \times \frac{1}{100}\right)$

□ D. Seven hundred twenty-one and three hundred sixty-nine thousandths

4. Place the appropriate comparison symbol <, =, > in the box.

$$(4 \times 10) + (6 \times 1) + \left(8 \times \frac{1}{10}\right) + \left(4 \times \frac{1}{100}\right) \square (4 \times 10) + (6 \times 1) + \left(8 \times \frac{1}{10}\right) + \left(3 \times \frac{1}{100}\right) + \left(9 \times \frac{1}{1000}\right)$$

5. Enter your answer in the box.

$483 \times 57 =$ [ ]

6. Enter your answer in the box.

$286 \times 1{,}769 =$ [ ]

7. Enter your answer in the box.

$7.8 \times 0.1 =$ [ ]

$7.8 \div 0.1 =$ [ ]

8. Enter your answer in the box.

$3{,}836 \div 28 =$ [ ]

9. Enter your answer in the box.

$7.68 + 26.37 =$ [ ]

10. Enter your answer in the box.

$48.52 - 29.78 =$ [ ]

11. Enter your answer in the box.

$3.4 \times 1.6 =$ [ ]

12. Enter your answer in the box.

$5.2 \div 0.4 =$ [ ]

13. Enter your answer in the box.

$\frac{5}{12} - \frac{1}{3} =$ [ ]

14. Enter your answer in the box.

$$\frac{2}{3}+\frac{1}{4}+\frac{3}{8}= \boxed{}$$

15. Enter your answer in the box.

$$\frac{3}{5}+\frac{1}{2}-\frac{3}{4}= \boxed{}$$

16. Enter your answer in the box.

$$4\frac{1}{8}+6\frac{2}{5}= \boxed{}$$

17. Kaitlyn subtracted the fractions below:

$$7\frac{5}{12}-2\frac{1}{3}= \boxed{}$$

Which **two** responses show how Kaitlyn solved the problem?

☐ A. Kaitlyn subtracted the whole numbers 7 – 2 and $\frac{5}{12}-\frac{1}{3}$ to arrive at an answer of $5\frac{4}{9}$.

☐ B. Kaitlyn subtracted the whole numbers 7 – 2 and $\frac{15}{36}-\frac{12}{36}$ to arrive at an answer of $5\frac{3}{36}$.

☐ C. Kaitlyn subtracted the whole numbers 7 – 2 and $\frac{5}{12}-\frac{4}{12}$ to arrive at an answer of $5\frac{1}{12}$.

☐ D. Kaitlyn subtracted the whole numbers 7 – 2 and $\frac{5}{12}-\frac{1}{12}$ to arrive at an answer of $5\frac{4}{12}$.

18. Your classmate arrived at the following solution for the problem below:

$$\frac{2}{5}+\frac{2}{4}=\frac{4}{9}$$

Which **one** of the following answers provides the best reason why the equation is incorrect?

○ A. $\frac{4}{9}<\frac{2}{5}$

○ B. $\frac{4}{9}<\frac{2}{4}$

○ C. $\frac{4}{9}=\frac{2}{4}$

○ D. $\frac{4}{9}<\frac{5}{2}$

19. Jocelyn needs to make $9\frac{3}{4}$ pounds of cookies for three math classes before Thursday.

- On Monday she baked $2\frac{5}{6}$ pounds of cookies.
- On Tuesday she baked $3\frac{3}{8}$ pounds of cookies.

Part A: What is the difference between the amount of cookies baked on the two days? Enter your answer in the space provided.

Part B: Tuesday afternoon the principal asks Jocelyn to make an additional $2\frac{1}{3}$ pounds of cookies for a fourth math class. How many remaining pounds of cookies does Jocelyn need to bake before Thursday. Enter your answer in the space provided.

20. Which **one** of the following answers provides the best explanation of the fraction below:

$$\frac{2}{5}$$

○ A. Five chocolate candy bars divided equally among two friends.

○ B. Two chocolate candy bars divided equally among five friends.

○ C. One chocolate candy bar is divided equally among two friends and the second chocolate candy bar is divided equally among three friends.

○ D. Two chocolate candy bars divided equally among four friends and one friend gets none.

21. Fifteen students sold an equal amount of candy for the school candy sale. The order was delivered in 125 boxes.

Part A: How many boxes of candy should each person get? Enter you answer in the space provided.

Part B: Between what two whole numbers does your answer lie? Enter your answer in the space provided.

________ and ________

22. In gym, you were required to run $\frac{2}{5}$ of a mile every day, for five days.

Select the equation that correctly identifies how many miles you ran in 5 days.

○ A. $\frac{2}{5} \div 5 = 2 \div 5 \times 5 = \frac{2}{25}$ miles

○ B. $\frac{2}{5} \times 5 = 2 \times 5 \div 5 = \frac{10}{5} = 2$ miles

○ C. $\frac{5}{2} \times 5 = 5 \times 5 \div 2 = \frac{25}{2} = 12\frac{1}{2}$ miles

○ D. $\frac{2}{5} + 5 = \frac{7}{5} = 1\frac{2}{5}$ miles

23. The fifth-grade class went to Dorney Park. That day $\frac{5}{8}$ of the class rode roller coasters. Later in the day $\frac{2}{3}$ of the roller coaster riders felt sick. What fraction of the class was ill from riding roller coasters? Write your answer in the box below.

24. Joshua's paper showed the following answer:

$$\frac{2}{3} \times \frac{2}{5} = \frac{4}{15}$$

Use a rectangular area model drawing to justify his answer. Show your work in the space provided.

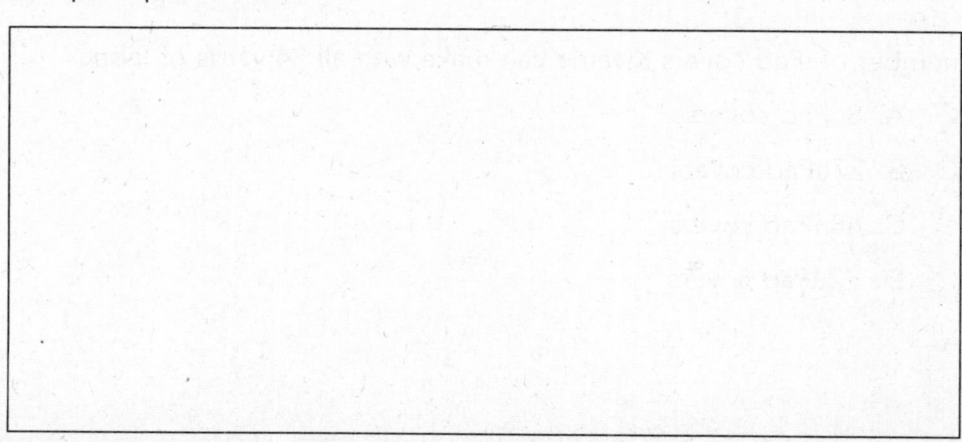

25. Select a phrase to correctly fill the blank of each sentence.

GREATER THAN   LESS THAN

The product of $\frac{1}{4}$ and 3 is ______________________________ 3.

The product of $2\frac{2}{3}$ and 7 is ______________________________ 7.

The product of $\frac{2}{5}$ and $\frac{11}{2}$ is ______________________________ $\frac{11}{2}$.

26. Kwan read $\frac{2}{3}$ of a novel that was $\frac{7}{8}$ inch thick. How much of the book (in inches) did he read? Write your answer in the box below.

27. Satesh makes $9 an hour working at Game Stop. Pay day is next week. Due to the Thanksgiving weekend, he worked $32\frac{2}{3}$ hours. How much will his check be next week? Write your answer in the box below.

28. Kwame uses felt fabric to make iPad covers. Kwame has 24 yards of felt fabric. He uses $\frac{1}{3}$ yard of felt fabric to make each cover. What is the total number of iPad covers Kwame can make with all 24 yards of fabric?

- ○ A. 8 iPad covers
- ○ B. 27 iPad covers
- ○ C. 48 iPad covers
- ○ D. 72 iPad covers

29. Complete the conversion by filling in the box with the correct number below:

0.05	500	0.5	5000	50

A. 5 cm = ☐ m

B. 0.5 km = ☐ m

C. 0.05 km = ☐ cm

30. The school cafeteria made 12 gallons of punch for the fifth-grade holiday party. You had 8 ounces of punch with your lunch, when Ms. Cathy, the cafeteria aide, recognized that she had accidently served 3 quarts of punch at lunch, including your glass.

Part A: How much punch is left for the party? Write your answer in the box below.

☐

Part B: Each student is allowed to have 1.5 pints of punch at the party. How many students will the remaining punch serve? Write your answer in the box below.

☐

31. You are doing a science experiment and have collected the following data for the height your bean plants have grown over 30 days.

$\frac{1}{2}$ inch	$\frac{1}{4}$ inch	$\frac{1}{4}$ inch
$\frac{1}{2}$ inch	$\frac{1}{2}$ inch	$\frac{1}{8}$ inch
$\frac{1}{4}$ inch	$\frac{1}{8}$ inch	$\frac{1}{8}$ inch
$\frac{3}{4}$ inch	$\frac{1}{2}$ inch	$\frac{1}{2}$ inch

Part A: Use the number line below to create a line plot to display the data.

0 ——————————————— 1

Part B: What is the difference in growth between the plant that grew the most and a plant that grew the least? Write your answer in the box below.

[ ]

32. The rectangular prism shown is made from cubes. Each cube is 1 cubic centimeter.

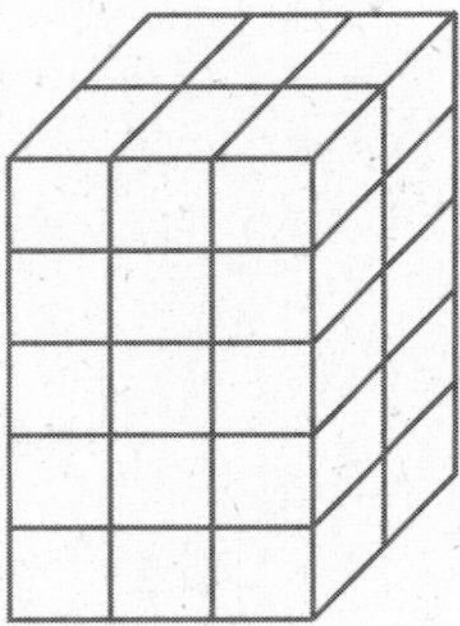

What is the volume of the rectangular prism shown? Write your answer in the box below.

[cubic cm]

33. A toy chest has a height of 36 inches and a base that is 24 inches long and 20 inches wide. What is the volume of the toy chest in cubic inches. Write your answer in the box.

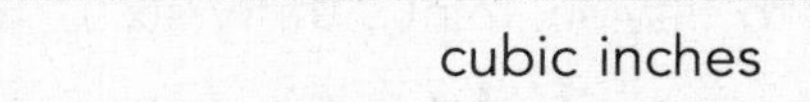

34. The park has two L-shaped community swimming pools, pool A and pool B. Each pool has a kiddie section and an adult section. The volume of the kiddie section of pool A is 192 cubic feet. The volume of the adult section for pool A is 12,996 cubic feet. What is the total volume, in cubic feet, of pool A?

Part A:

- ○ A. 12,804 cubic feet
- ○ B. 68 cubic feet
- ○ C. 13,188 cubic feet
- ○ D. 6,594 cubic feet

Part B: Pool B has the same volume as pool A. The volume of the kiddie section of pool B is 252 cubic feet. What is the volume of the adult section, in cubic feet, for pool B? Write you answer in the box below.

[cubic feet]

35. Enter your answer in the box.

$5 \times (6 + 2) \div 4 =$ [ ]

36. Select the correct expression for the written statement below.

6 more than 2 times a number x

- ○ A. $6x + 2$
- ○ B. $2x - 6$
- ○ C. $6 + 2$
- ○ D. $6 + 2x$

37. Select the correct written statement for the expression below.

$$7 \times (52 - 36)$$

○ A. Seven added to fifty-six minus thirty-six

○ B. Seven times the difference between 52 and 36

○ C. Seven times fifty-two minus thirty-six

○ D. Seven more than fifty-two minus thirty-six

38. Which **two** statements about the corresponding terms in both Pattern A and Pattern B are always true?

Pattern A: 0, 1, 2, 3, 4, 5

Pattern B: 0, 3, 6, 9, 12, 15

☐ A. Each term in Pattern B is 3 times the corresponding term in Pattern A.

☐ B. Each term in Pattern A is 3 times the corresponding term in Pattern B.

☐ C. Each term in Pattern A is 2 less than the corresponding term in Pattern B.

☐ D. Each term in Pattern A is $\frac{1}{3}$ of the corresponding term in Pattern B.

39. Graph points *A*, *B*, and *C* on the coordinate plane. Point *A* should be located at (3, 2), point *B* should be located at (2, 3), and point *C* should be located at (2, 7). Graph all three points.

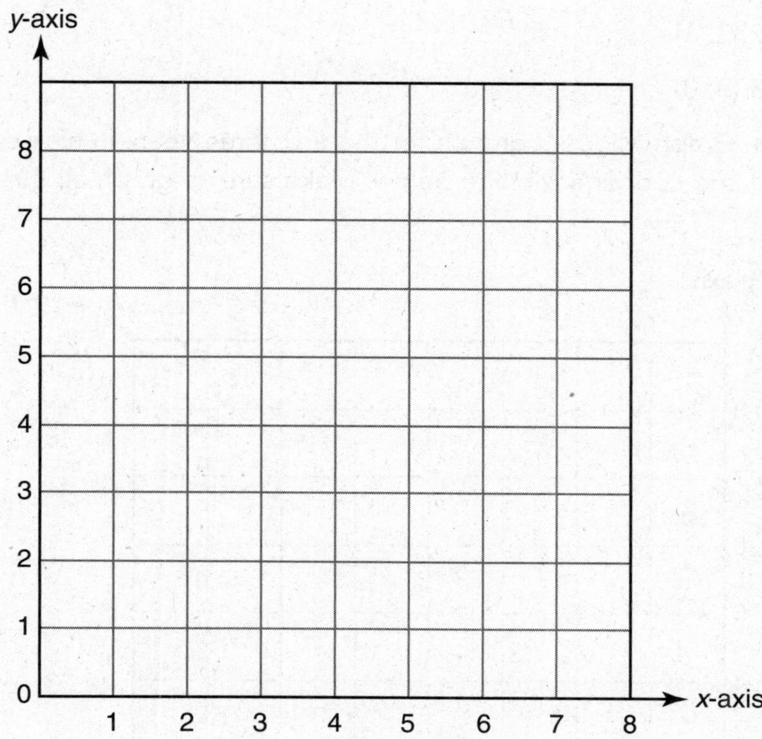

40. Miguel has been keeping track of his rate of growth since moving to his new home 4 years ago. Each coordinate pair represents the year lived in his home and the number of inches he grew that year.

- Year 1 (1, 4)
- Year 2 (2, 1)
- Year 3 (3, 3)

Part A: Graph Miguel's growth for the first three years in his new home on the coordinate plane below. Make sure to graph all three points.

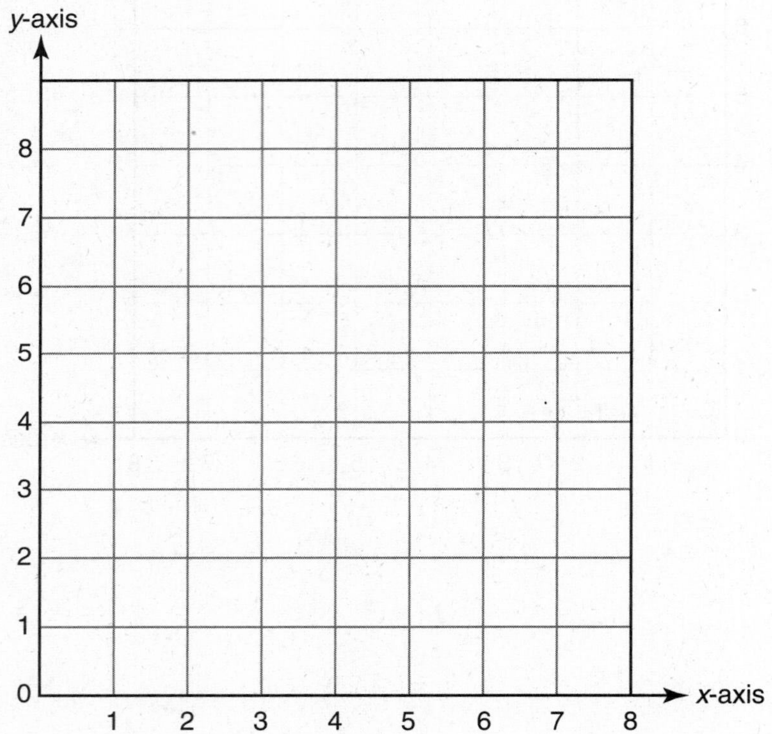

Part B: The fourth year in his home, Miguel grew two times the number of inches in years 1 and 2 combined.

What is the coordinate pair that represents Miguel's height in year 4?

○ A. (3, 10)

○ B. (4, 8)

○ C. (4, 10)

○ D. (5, 8)

41. Which explanation about figures is **not** correct?

- ○ A. A square is a trapezoid because it has exactly one pair of parallel sides.
- ○ B. A square is a parallelogram because it has two pairs of parallel sides.
- ○ C. A square is a rectangle because it has four right angles, and the opposite sides are equal and parallel.
- ○ D. A square is a rhombus because it has four equal sides and two pairs of parallel sides.

42. Use your knowledge of classifying quadrilaterals to answer the questions below:

	YES	NO
All rectangles are squares.		
All quadrilaterals are trapezoids.		
All rhombuses are parallelograms.		
All squares are rhombuses.		
All parallelograms are quadrilaterals.		
All rectangles are rhombuses.		

Diagnostic Assessment Answer Key

Question	Answer	CCSS	Detailed Explanation
1	B, C	5.NBT.1	A digit in the tenths place represents a value 10 times greater than the same digit in the hundredths place. A digit in the hundredths place represents a value that is $\frac{1}{10}$ the value of the same digit in tenths place.
2	C	5.NBT.2	$30{,}000 = 3 \times 10{,}000 = 3 \times 10^4$ $10{,}000 = 10 \times 10 \times 10 \times 10$ 10 is multiplied by itself 4 times. The exponent 4 tells how many times to use 10 as a factor, known as a power of 10. Note that the power of 10 has the same number of zeros as the exponent.
3	B, D	5.NBT.3a	721.369 is written in word form and expanded form.
4	$>$	5.NBT.3b	In comparing the numbers the values are the same until we reach the hundredths place. The value of $4 \times \frac{1}{100}$ in the number 46.84 is greater than the value of $3 \times \frac{1}{100}$ in the number 46.839. Since we do not have a digit in the thousandths place of the number 46.84 we do not have to consider the digit in the thousandths place of the number 46.839.
5	27,531	5.NBT.5	Using the algorithm, you arrive at the answer. $\begin{array}{r} {}^{4\,1}_{5\,2} \\ 483 \\ \times\ 57 \\ \hline 3381 \\ +\ 24150 \\ \hline 27{,}531 \end{array}$

Question	Answer	CCSS	Detailed Explanation
6	505,934	5.NBT.5	Using the algorithm, you arrive at the answer. $\begin{array}{r} 1769 \\ \times\ 286 \\ \hline 10614 \\ 141520 \\ +\ 353800 \\ \hline 505{,}934 \end{array}$
7	7.8 × 0.1 = 0.78 7.8 ÷ 0.1 = 78	5.NBT. int.1	Task does not require you to multiply but to understand place value system. $7.8 \times 0.1 = 7.8 \times \frac{1}{10}$ must be an answer less than 1. $7.8 \div 0.1 = 7.8 \div \frac{1}{10}$ must be an answer greater than than 1.
8	137	5.NBT.6	Using place value understanding, we can divide 3,836 by 28. 28)3836 2800	100 1036 280	10 756 280	10 476 280	10 196 56	2 140 56	2 84 56	2 28 28	1 0 100 + 10 + 10 +10 + 2 + 2 + 2 + 1 = 137

Question	Answer	CCSS	Detailed Explanation
9	34.05	5.NBT.7-1	7.0 + 0.6 + 0.08 + 26.0 + 0.3 + 0.07 Write numbers in expanded form. When adding or subtracting decimals line up the decimal points. 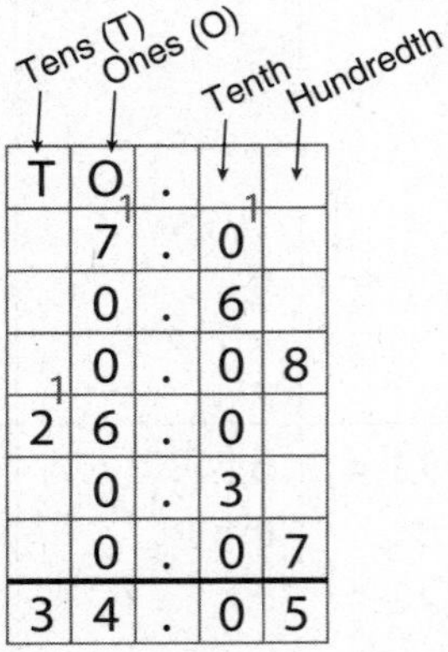 Tens (T) Ones (O) Tenth Hundredth T O . 7 . 0 0 . 6 0 . 0 8 2 6 . 0 0 . 3 0 . 0 7 3 4 . 0 5
10	18.74	5.NBT.7-2	When adding or subtracting decimals, line up the decimal points. Remember the strategy of regrouping to subtract. 48.52 − 29.78 18.74
11	5.44	5.NBT.7-3	Using an area model to multiply decimals, we have 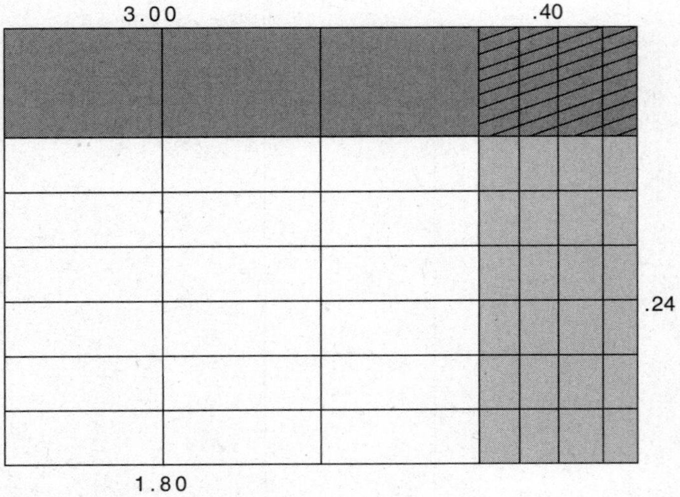 3.00 .40 .24 1.80 3.4 ×1.6 .24 1.80 .40 3.00 5.44

Question	Answer	CCSS	Detailed Explanation
12	13	5.NBT.7-4	Remember $\frac{10}{10}=1$ whole. You have 52 tenths of some object that needs to be divided into 4 tenth sections. How many groups of 4 tenths do you have? Skip counting by 4, you arrive at the answer of 13. 4 tenths, 8 tenths, 12 tenths, 16 tenths, 20 tenths, 24, 28, 32, 36, 40, 42, 48, 52 tenths
13	$\frac{3}{36}$ or $\frac{1}{12}$	5.NF.1	Remember the key is to find the common denominator by multiplying the denominators $\frac{15}{36}-\frac{12}{36}=\frac{3}{36}=\frac{1}{12}$ or you can use equivalent fractions: $\frac{5}{\textcircled{12}}$: $\frac{1}{3}$: $\frac{2}{6}, \frac{3}{9}, \frac{4}{\textcircled{12}}$ $\frac{5}{12}-\frac{4}{12}=\frac{1}{12}$
14	$\frac{31}{24}$ or $1\frac{7}{24}$	5.NF.1-2	Find common denominators or use equivalent fractions.
15	$\frac{7}{20}$	5.NF1-3	Find common denominators or use equivalent fractions.
16	$10\frac{21}{40}$	5.NF.1-4	Find common denominators or use equivalent fractions. $4+6+\frac{5}{40}+\frac{16}{40}=10\frac{21}{40}$
17	B, C	5.NF.1-5	Find common denominators or use equivalent fractions. $5\frac{3}{36}$ or $5\frac{1}{12}$
18	B	5.NF.2-2	Recognize that $\frac{2}{4}$ represents $\frac{1}{2}$ and $\frac{4}{9}$ is less than $\frac{1}{2}$.
19	Part A: $\frac{26}{48}$ or $\frac{13}{24}$ lb Part B: $5\frac{21}{24}$ or $5\frac{7}{8}$	5.NF.2-1	Use your knowledge of finding a common denominator to subtract the fractions. Use your knowledge of finding the common denominator to add and subtract fractions.

Question	Answer	CCSS	Detailed Explanation
20	B	5.NF.3-1	$\frac{2}{5} = 2 \div 5$ Two chocolate bars divided equally among five friends.
21	Part A: $8\frac{1}{3}$ Part B: 8 and 9	5.NF.3-2	125 boxes divided equally among 15 people $\frac{125}{15} = 125 \div 15 = 8\frac{1}{3}$
22	B	5.NF.4a-1	$\frac{a}{b} \times q = \frac{aq}{b}$ Think of the equation in this way. Multiplication represents repeated addition. So you ran $\frac{2}{5} + \frac{2}{5} + \frac{2}{5} + \frac{2}{5} + \frac{2}{5} = \frac{10}{5}$ the denominator remains the same. This is equivalent to $\frac{2}{5} \times 5 = \frac{2 \times 5}{5} = \frac{10}{5} = 2$ miles.
23	$\frac{10}{24}$ or $\frac{5}{12}$	5.NF.4a-2	$\frac{a}{b} \times \frac{c}{d} = \frac{ac}{bd}$ $\frac{5}{8} \times \frac{2}{3} = \frac{10}{24} = \frac{5}{12}$
24	See Detailed Explanation.	5.NF.4b-1	Imagine you have a 3×5 array that has $\frac{2}{3}$ of the rows shaded and $\frac{2}{5}$ of the columns shaded. The overlapping area represents $\frac{4}{15}$.
25	A. Less Than B. Greater Than C. Less Than	5.NF.5a	A. You are multiplying a whole number by a number less than one, so the product must be less than 3. B. You are multiplying a whole number by a number greater than one, so the product must be more than 7. C. You are multiplying more than a whole number by a number less than one, so the product must be less than $\frac{11}{2}$.

<table>
<tr><th>Question</th><th>Answer</th><th>CCSS</th><th>Detailed Explanation</th></tr>
<tr><td>26</td><td>$\frac{14}{24}$ or $\frac{7}{12}$</td><td>5.NF.6-1</td><td>See visual image of $\frac{2}{3} \times \frac{7}{8} = \frac{14}{24}$.</td></tr>
<tr><td>27</td><td>$294</td><td>5.NF.6-2</td><td>One approach to solving the problem
$32\frac{2}{3} \times \$9 = ?$
$\frac{98}{3} \times \$9 = \frac{98 \times 9}{3} = \frac{882}{3} = \$294$</td></tr>
<tr><td>28</td><td>D</td><td>5.NF.7c</td><td>This problem is similar to question 12.
One approach:
You have 24 yards of fabric that need to be divided into $\frac{1}{3}$ yard pieces. How many groups of $\frac{1}{3}$ do you have? Note: $\frac{3}{3} = 1$ whole or 1 yard. Each yard will make 3 covers. 24 yards × 3 = 72 covers.
1 whole = $\frac{3}{3}$</td></tr>
<tr><td>29</td><td>A. 0.05 m B. 500 m C. 5,000 cm</td><td>5.MD.1</td><td>Remember to convert to a smaller unit, move the decimal point to the right or multiply by a power of 10. |	km	hm	dk	m	dm	cm	mm
A				0.05	0.5	5	
B	0.5	5	50	500			
C	0.05	0.5	5	50	500	5,000	| To convert to a larger unit, move the decimal point to the left and divide by a power of 10.</td></tr>
</table>

Question	Answer	CCSS	Detailed Explanation
30	Part A: 1,440 ounces or 11.25 gallons Part B: 60 students	5.MD.1-2	1 gallon = 128 ounces 1 quart = 2 pints 1 pint = 16 ounces Part A: 12 gallons × 128 ounces = 1,536 ounces 3 quarts = 6 pints = 96 ounces 1,536 ounces – 96 ounces = 1,440 ounces remaining Part B: 1.5 pints = 24 ounces 1,440 ounces ÷ 24 ounces = 60 students
31	Part A: See Detailed Explanation. Part B: $\frac{5}{8}$ in	5.MD.2-2	Part A: Line Plot 0: (none); $\frac{1}{8}$: x x x; $\frac{1}{4}$: x x x; $\frac{1}{2}$: x x x x x; $\frac{3}{4}$: x; 1: (none) Part B: Use knowledge of subtracting fractions to arrive at your answer.
32	30 cubic cm	5.MD.4	Count the blocks. 30 blocks means volume is 30 cubic centimeters.
33	17,280 cubic inches	5.MD.5b	36 inches × 24 inches × 20 inches
34	Part A: C Part B: 12,936 cubic feet	5.MD.5c	192 cubic feet + 12,996 cubic feet = 13,188 cubic feet 13,188 cubic feet – 252 cubic feet = 12,936 cubic feet
35	10	5.OA.1	To evaluate this math sentence containing brackets, proceed in this order: 1. Parentheses are done first 2. Multiplication and division are done as they are encountered from left to right. Ignoring the grouping symbols and approaching the problem any other way will lead to a totally different answer. $5 \times (6 + 2) \div 4 = 5 \times 8 \div 4 = 40 \div 4 = 10$

Question	Answer	CCSS	Detailed Explanation
36	D	5.OA.2-1	$6x + 2$ = six times a number x and 2 more $2x - 6$ = 2 times a number x less six $6 + 2$ = 6 more than two $6 + 2x$ = 6 more than 2 times a number x
37	B	5.OA.2-2	Same premise as question 35. However, the braces identify that the difference between 52 – 36 should be the value multiplied by 7.
38	A, D	5.OA.3	Pattern A is $\frac{1}{3}$ the values in Pattern B, and Pattern B represents 3 times the values in Pattern A.
39	See Detailed Explanation.	5.G.1	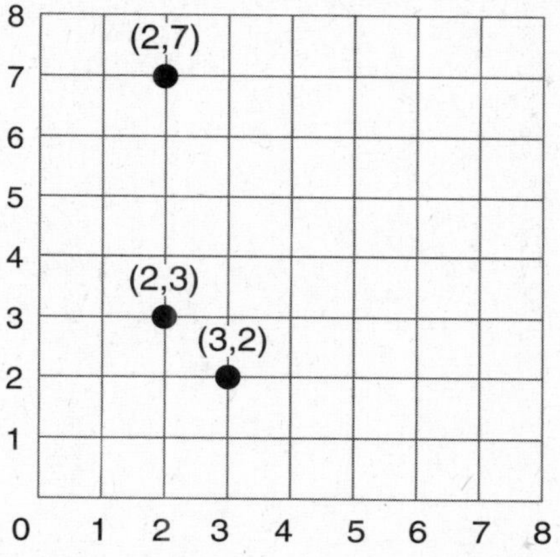 Graph: points plotted at (2,7), (2,3), (3,2); axes 0–8
40	Part A: See Detailed Explanation. Part B: C		A. 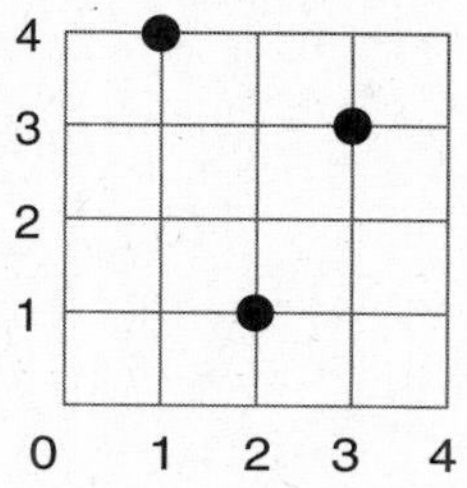 Graph: points plotted at (1,4), (3,3), (2,1); axes 0–4 B. Coordinate pair year 4 = (4, 2(4 + 1)) = (4, 10)
41	A	5.G.3	A square has exactly two pairs of parallel sides and thus is not a trapezoid.
42	See Detailed Explanation.	5.G.4	See table below

	YES	NO
All rectangles are squares.		X
All quadrilaterals are trapezoids.		X
All rhombuses are parallelograms.	X	
All squares are rhombuses.	X	
All parallelograms are quadrilaterals.	X	
All rectangles are rhombuses.		X

Number and Operations in Base Ten

CHAPTER 3

I recently had the opportunity to go to the movies with my son over the holidays to see a film that dealt with ordinary genius kids turned superheroes. They actually used their genius inventions as the source of their superpowers. Okay, so what does all this have to do with math and more so Numbers and Operations in Base Ten? Not a lot. However, I'm somewhat of a numbers geek. So, if you asked me to think about a number being a superhero and having superpowers, well I would have to say that would be the number 10!

Sounds crazy I know, but imagine this! Base 10, from the planet of Place Value System, has superpowers that can increase any item, object, or thing you have by 10!!! Okay, what kid wouldn't love that! More so, how simple is that math? You have one Xbox One console, until you call on Superhero Base 10. He shows up and CAPOW!!! Now how many Xbox One consoles do you have? 10! You have 20 Lego sets, you call on Base 10 and now you have? You are catching on, right! 200!

Now here's the downside for kids and probably an upside for parents. Can't get you to clean up your toys or pick stuff up off the floor? So, just call on Superhero Base 10! You will learn that he has the power to decrease things by 10, too! So you left those 30 baby dolls scattered all over the floor. Capow!!!! How many do you have now? Yep, only three!!! Shouldn't be too hard to clean those up!

Back to being the serious numbers person now! You will learn to love Base 10 and why he or she is from the planet of Place Value System! His or her powers provide you with the ability to perform the easiest form of multiplication, and you will also begin to understand this power as a form of division, as you learn to connect this to the place value system. What is best is that once you understand Base 10, it becomes a mental math strategy that does not require the use of a calculator. What am I talking about?

Just push to learn more about Number and Operations in Base Ten and you will see!

VISUALIZE THE MATH

1. Look at the models below. What is the value of each picture? Write your answer in the box provided below. The first answer has been provided for you.

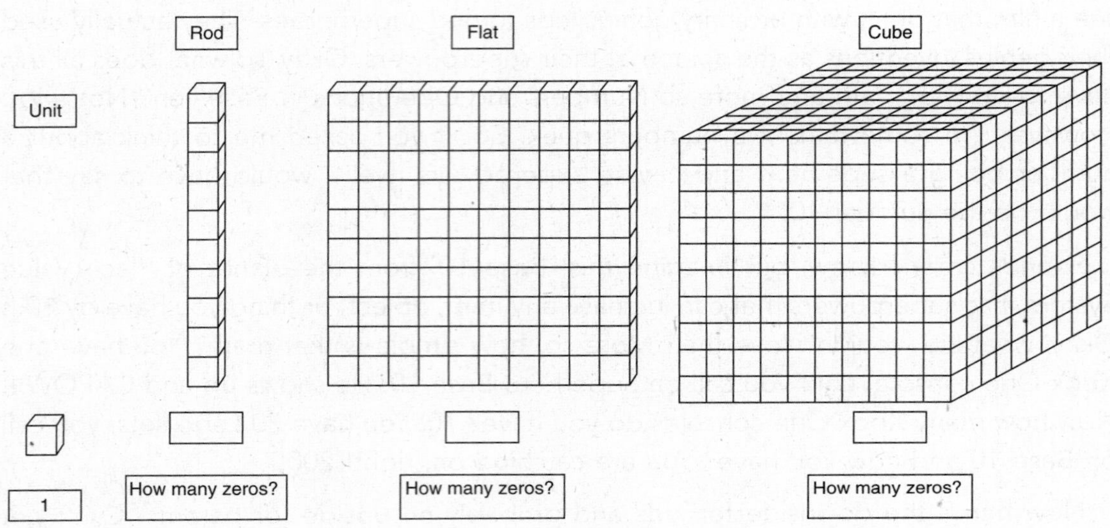

2. Look at your answers above; what pattern do you see?

_______________ _______________ _______________

3. Use the pattern from above to write the following next three numbers for your pattern and explain how you know you are correct.

_______________ _______________ _______________

How do you know? Write you thoughts here.

Answers

1.

Rod	10	1 zero
Flat	100	2 zeros
Cube	1,000	3 zeros

2. I keep adding a zero to the next number, the number is 10 times larger, or any similar response.

3. 10,000 100,000 1,000,000
   I keep following the pattern of adding zeros so that I have 4 zeros, then 5 zeros, and then 6 zeros.

Powers of Ten

THE STANDARD

CCSS 5.NBT.2. Explain patterns in the number of zeros of the product when multiplying a number by powers of 10, and explain patterns in the placement of the decimal point when a decimal is multiplied or divided by a power of 10. Use whole-number exponents to denote powers of 10.

What Does This Mean?

You should notice that every time you multiply or divide a number by a power of 10 a special pattern is created in terms of the number of zeros in the answer or the decimal point placement in the answer. You learn that powers of 10 can be written using an exponent, which relates to the pattern of zeros or decimal placement in your answer.

What Do We Mean When We Say Power of 10?

A power means you are multiplying the number repeatedly by itself.

$$10 \times 10 \times 10 \times 10 = 10^4 = 10{,}000 = \text{ten thousand}$$

In the previous example, the repeated factor is 10. The repeated factor is called your **base**. The base is repeated four times, so your **exponent** is 4. Notice, the number of zeros in your product is four. You can connect that your product will contain the same number of zeros as the exponent.

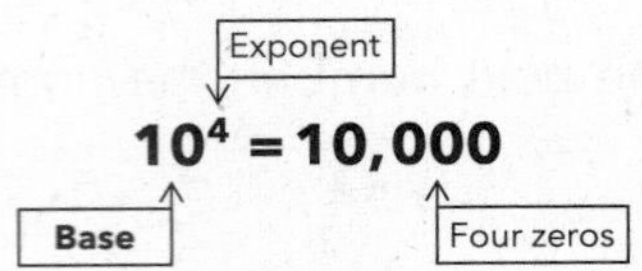

Power of 10	Read	Factors	Word Name
$10^0 = 1$	The zeroth power of ten.	1	One
$10^1 = 10 \times 1 = 10$	The first power of ten.	10	Ten
$10^2 = 10 \times 10 = 100$	The second power of ten.	10×10	Hundred
$10^3 = 10 \times 100 = 1{,}000$	The third power of ten.	$10 \times 10 \times 10$	Thousand
$10^4 = 10 \times 1{,}000 = 10{,}000$	The fourth power of ten.	$10 \times 10 \times 10 \times 10$	Ten Thousand

What Is the Pattern?

By now you have noticed that every time you multiply by another power of ten you add another zero. That is absolutely correct, but do you know why? Let's look a little closer.

Example

$$10^0 = 1$$

$$10^1 = 10$$

What you do not see above is the decimal point behind the 1 … right there.

In reality,

$$10^0 = 1 = 1.0$$

What Is the Connection to the Placement of the Decimal Point?

Each time you multiply by another power of ten, you are actually moving the decimal point in your original number, to the right, the same number of places as the exponent.

Original Number	10^1	10^2	10^3
$10^0 = 1$	1 × 10	1 × 100	1 × 1,000
1.0	10.	100.	1000.
Move decimal 0 places	Move decimal 1 place	Move decimal 2 places	Move decimal 3 places

Each time you divide by another power of ten, you are actually moving the decimal point in your original number, to the left, the same number of places as the exponent.

Original Number	10^1	10^2	10^3
$10^0 = 1$	1 ÷ 10	1 ÷ 100	1 ÷ 1,000
1.0	0.1	0.01	0.001
Move decimal 0 places	Move decimal 1 place	Move decimal 2 places	Move decimal 3 places

Multiplying or Dividing Whole Numbers and Decimals by a Power of 10

Now that you recognize the pattern and have an understanding of the power of 10, let's look at what happens when you multiply or divide numbers by a power of 10.

Example: Multiplying by a Power of 10

Whole Number	Decimals
$7 \times 10^0 = 7$	$7.125 \times 10^0 = 7.125$
$7 \times 10^1 = 70$	$7.125 \times 10^1 = 71.25$
$7 \times 10^2 = 700$	$7.125 \times 10^2 = 712.5$
$7 \times 10^3 = 7{,}000$	$7.125 \times 10^3 = 7{,}125$

Example: Dividing by a Power of 10

Whole Number	Decimals
$175 \div 10^0 = 175$	$87.3 \div 10^0 = 87.3$
$175 \div 10^1 = 17.5$	$87.3 \div 10^1 = 8.73$
$175 \div 10^2 = 1.75$	$87.3 \div 10^2 = 0.873$
$175 \div 10^3 = 0.175$	$87.3 \div 10^3 = 0.0873$

HELPFUL HINT

When you multiply by a power of 10, the number gets larger as the decimal point is moved to the right. When you divide by a power of 10, the number gets smaller as the decimal is moved to the left.

PRACTICE—CHECK UNDERSTANDING: Powers of Ten

Complete the table below.

	Value	Factors	Exponent Form
1.			10^6
2.	100,000		
3.		$10 \times 10 \times 10$	
4.	10,000		

Use your knowledge of the powers of 10 to solve the problems below.

5. Which expression shows the ninth power of 10?

○ A. 9×10^1

○ B. 10^9

○ C. 10×9^1

○ D. 10^{90}

6. $568 \div 10^0 =$ ________

$568 \div 10^1 =$ ________

$568 \div 10^2 =$ ________

$568 \div 10^3 =$ ________

7. Write the number 9,356 as a decimal number multiplied by a power of 10 in the box below.

[ ]

8. $2.649 \times 10^0 =$ ________

$2.649 \times 10^1 =$ ________

$2.649 \times 10^2 =$ ________

$2.649 \times 10^3 =$ ________

9. Select the two statements that are equal to 0.00524×10^4.

□ A. $0.00524 \times 10 \times 10 \times 10$

□ B. 52.4

□ C. $0.00524 \times 10 \times 10 \times 10 \times 10$

□ D. 5.24

10. The distance from the earth to the moon is about 240,000 miles. Write the distance as a whole number multiplied by a power of 10 in the box below.

[ ]

11. Select the three expressions that equal 275,000.

□ A. 275×10^2

□ B. 2.75×10^5

□ C. 27.5×10^4

□ D. 275×10^3

12. $1{,}275 \div 10^0 =$ ________

$1{,}275 \div 10^1 =$ ________

$1{,}275 \div 10^2 =$ ________

$1{,}275 \div 10^3 =$ ________

13. Write the number 76,500 as a whole number multiplied by a power of 10 in the box below.

[ ]

14. $0.49 \times 10^0 =$ ________

$0.49 \times 10^1 =$ ________

$0.49 \times 10^2 =$ ________

$0.49 \times 10^3 =$ ________

Answers (pages 42–43)

	Value	Factors	Exponent Form
1	1,000,000	$10 \times 10 \times 10 \times 10 \times 10 \times 10$	10^6
2	100,000	$10 \times 10 \times 10 \times 10 \times 10$	10^5
3	1,000	$10 \times 10 \times 10$	10^3
4	10,000	$10 \times 10 \times 10 \times 10$	10^4

Question	Answer	Detailed Explanation
5	B	The base is 10. The exponent tells you how many times 10 is multiplied by itself. The ninth power = exponent 9.
6	568 56.8 5.68 0.568	Remember that when dividing by a power of 10, the decimal point is moved to the left the same number of places as the exponent.
7	935.6×10^1 93.56×10^2 9.356×10^3 0.9356×10^4	Any response that shows an understanding of multiplying a decimal by a power of ten.
8	2.649 26.49 264.9 2,649.0	Remember that when multiplying by a power of 10, the decimal point is moved to the right the same number of places as the exponent.
9	B, C	
10	24×10^4	or 240×10^3
11	B, C, D	275,000 = 2.75×10^5 = 2.75000 = 275,000 27.5×10^4 = 27.5000 = 275,000 275×10^3 = 275.000 = 275,000
12	1,275 127.5 12.75 1.275	

Question	Answer	Detailed Explanation
13	765×10^2	or $7,650 \times 10^1$
14	0.49 4.9 49 490	Remember that when multiplying by a power of 10, the decimal point is moved to the right the same number of places as the exponent.

Place Value

Now that you are familiar with the powers of 10, it is time to discuss the relationship between base-ten and the place value system. You discovered that when you multiplied by a power of 10 the number got larger as you moved from the ones, to the tens, to the hundreds place, and so on. However, when you divided by a power of 10 depending upon the value of the digit the number got smaller and smaller and smaller based on the position of the decimal point. So now, let's look at what happens to the value of the same digit depending on its place in a number.

THE STANDARDS

CCSS 5.NBT.1. Recognize that in a multi-digit number, a digit in one place represents 10 times as much as it represents in the place to its right and $\frac{1}{10}$ of what it represents in the place to its left.

CCSS 5.NBT.3a. Read, write, decimals to thousandths.

A. Read and write decimals to thousandths using base-ten numerals, number names, and expanded form, e.g., $347.392 = 3 \times 100 + 4 \times 10 + 7 \times 1 + 3 \times \left(\frac{1}{10}\right) + 9 \times \left(\frac{1}{100}\right) + 2 \times \left(\frac{1}{1000}\right)$.

What Does This Mean?

You are building on your fourth-grade knowledge of looking at the relationship of the digits in multi-digit numbers and using decimal notation for fractions with denominators of 10 or 100. You will be introduced to decimal notation called the thousandths, meaning the fraction has a denominator of 1,000. You are also able to apply your understanding of the powers of 10 to the place value system. For example, you recognize that the place value to the left of 1 is 10 times greater in value, and that the place value to the right of 1 is 10 times less in value. You understand that the number less than 1 is called a decimal number and that the placement of the decimal point will affect the number's value. Additionally, you know at least three ways (base-ten numbers, number names, and expanded form) to read and write numbers that express the value of each digit.

Let's begin taking a deeper look into the place value system starting with whole numbers.

The Place Value System—Whole Numbers

What Do We Mean When We Say a Digit's Place Value?

Simply put, it is the value of a digit based upon its location in the number. Let's start by looking at the digit 7 and creating a whole number using only the digit 7.

777

Just based on using your understanding of the powers of 10 you know that the digits in the number above are 10 times more the digit to right and 10 times less the digit to the left.

However, what can you use to help you arrange the numbers so that you can visually see this? **A place value chart.**

Number is getting smaller →

	$\frac{1}{10}$ of the hundred millions place	$\frac{1}{10}$ of the ten millions place	$\frac{1}{10}$ of the one millions place	$\frac{1}{10}$ of the hundred thousands place	$\frac{1}{10}$ of the ten thousands place	$\frac{1}{10}$ of the thousands place	$\frac{1}{10}$ of the hundreds place	$\frac{1}{10}$ of the tens place
Millions Period			Thousands Period			Ones Period		
Hundreds	Tens	Ones	Hundreds	Tens	Ones	Hundreds	Tens	Ones
10 x the ten millions place	10 x the one millions place	10 x the hundred thousands place	10 x the ten thousands place	10 x the thousands place	10 x the hundreds place	10 x the tens place	10 x the ones place	
						7	7	7

← Number is getting larger

Looking at the place value chart, you can see that each of the digits above is named by a specific value based on its location on the chart. Knowing the place value of the digits helps you to read and write the number in standard form, number name, or word form; expanded form; and base-ten numerals.

The **number name** or **word form** of 777 is seven hundred seventy-seven. Each digit takes on the place value of the corresponding column. The value of the 7 in the hundreds column is 700, the value of the 7 in the tens column is 70, and the value of the 7 in the ones column is 7.

Knowing the number name or word form of 777 assists you in being able to express the number in **expanded form**. So now you can write seven hundred seventy-seven as **700 + 70 + 7**. Lastly, using your knowledge of the powers of 10 you can express the number using **base-ten numerals** as $(7 \times 100) + (7 \times 10) + (7 \times 1)$ or base-ten numerals using exponents $(7 \times 10^2) + (7 \times 10^1) + (7 \times 10^0)$.

Example

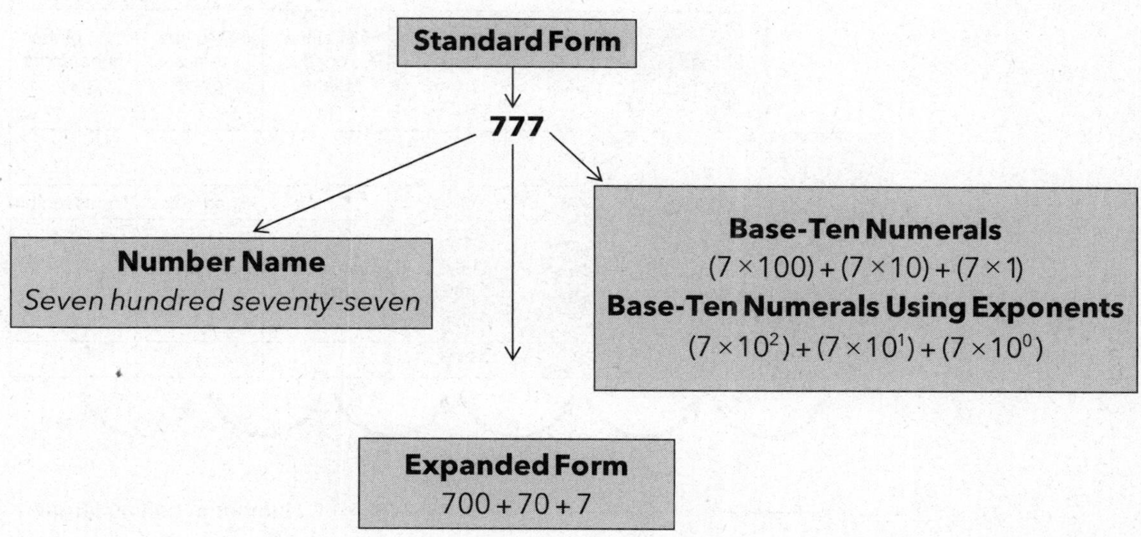

Next, you have to look at how this changes when you are talking about decimals.

The Place Value System—Decimal Numbers

What Do We Mean When We Say a Digit's Place Value for a Decimal Number?

In fourth grade, you began to identify that not every number is a whole number, and special notation, called decimal notation, is used for fractions with denominators of 10 or 100. So you should understand that a decimal point is used to separate the whole numbers on the left, from the decimal digits on the right. Simply put, because of that, the place value system extends to the right of the decimal point. Just as with whole numbers, the value of a decimal digit is based upon its location in the number. Let's start by looking at the digit 7 and creating a decimal number using only the digit 7.

7.777

Since you understand the powers of 10, you know that the decimal digits in the number above are 10 times more the digit to right, and 10 times less the digit to the left.

Again, you will use a **place value chart** to assist you in visually seeing this.

Number is getting smaller →

	$\frac{1}{10}$ of the thousands place	$\frac{1}{10}$ of the hundreds place	$\frac{1}{10}$ of the tens place	$\frac{1}{10}$ of the ones place	$\frac{1}{10}$ of the tenths place	$\frac{1}{10}$ of the hundredths place
	Ones Period					
Thousands	Hundreds	Tens	Ones	Tenths	Hundredths	Thousandths
10 x the hundreds place	10 x the tens place	10 x the ones place	10 x the tenths place	10 x the hundredths place	10 x the thousandths place	
			7	7	7	7

← Number is getting larger

Looking at the place value chart above, you can see that each of the digits is named by a specific value based on its location in the chart. Knowing the place value of the digits helps you to read and write the number in standard form, number name or word form; expanded form; and base-ten numerals.

The **number name** or **word form** of 7.777 is seven and seven hundred seventy-seven thousandths. Just as with whole numbers each digit takes on the place value of the corresponding column. The value of the seven in the ones column is seven, the value of the seven in the tenths column is seven tenths, the value of the seven in the hundredths column is seven hundredths, and the value of the seven in the thousandths column is seven thousandths.

KEY CONCEPT

Decimals are named by the place value of their final digit.

HELPFUL HINT

When reading and writing decimals use AND to represent the decimal point.

Knowing the number name or word form of 7.777 assists you in being able to express the number in **expanded form**. So now you can write seven and seven hundred seventy-seven thousandths as 7.0 + 0.7 + 0.07 + 0.007. Lastly, using your knowledge of the powers of 10, you can express the number using **base-ten numerals** as $(7 \times 1) + \left(7 \times \frac{1}{10}\right) + \left(7 \times \frac{1}{100}\right) + \left(7 \times \frac{1}{1000}\right)$ or dividing by base-ten numerals using exponents $(7 \div 10^0) + (7 \div 10^1) + (7 \div 10^2) + (7 \div 10^3)$.

Example

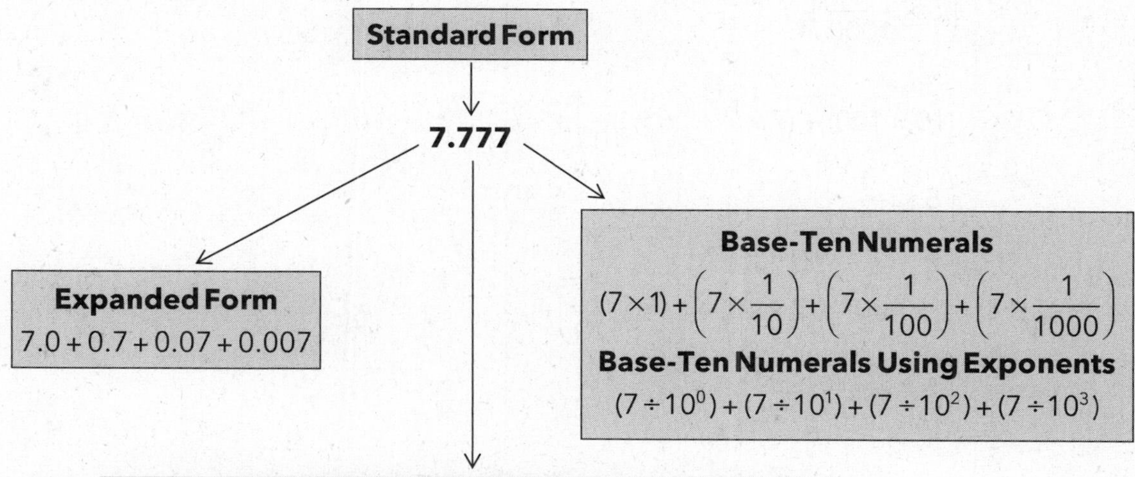

PRACTICE—CHECK UNDERSTANDING: Place Value

Use your knowledge of place value to solve the problems below.

1. Write 68,975 in expanded form in the box below.

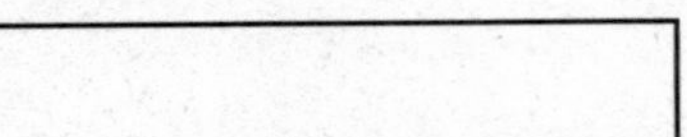

2. The number 107.065 can be expressed as:

○ A. $(1 \times 100) + (7 \times 1) + \left(6 \times \frac{1}{100}\right) + \left(5 \times \frac{1}{1000}\right)$

○ B. $(1 \times 100) + (7 \times 1) + \left(6 \times \frac{1}{10}\right) + \left(5 \times \frac{1}{1000}\right)$

○ C. $(10 \times 100) + (7 \times 1) + \left(6 \times \frac{1}{10}\right) + \left(5 \times \frac{1}{100}\right)$

○ D. $(1 \times 100) + (7 \times 10) + \left(6 \times \frac{1}{100}\right) + \left(5 \times \frac{1}{1000}\right)$

3. Select the answers that show the relationship between 0.007 and 0.07.

☐ A. 0.007 *is equal to* 0.07

☐ B. 0.07 *is* 10 *times* 0.007

☐ C. 0.007 *is* $\frac{1}{10}$ *of* 0.07

☐ D. 0.007 *is* 10 *times* 0.07

4. Given the numbers 324 and 32.4, explain the difference in place value between the 3 in each number.

5. Tyler expressed the number 480,730 as $(4 \times 10^5) + (8 \times 10^4) + (7 \times 10^3) + (3 \times 10^1)$.

Use your knowledge of base-ten numerals and place values to explain why Tyler's answer is incorrect.

____

____

____

____

____

Select the correct standard form value for the number names below.

6. Sixty-four and five hundred eighteen thousandths

- ○ A. 64.518
- ○ B. 64.0518
- ○ C. 64.815
- ○ D. 6,400.518

7. Ninety-one and eighty-four hundredths

- ○ A. 91.084
- ○ B. 19.8
- ○ C. 91.84
- ○ D. 91.0084

Use place value to determine the value of the digit.

8. 34,529.97
The 9 in the ones place is _______ times the value of the 9 in the tenths place.

____

9. 7,560.61
The 6 in the tens place is _______ times the value of the 6 in the tenths place.

____

10. 953.959
The 5 in the tens place is _______ times the value of the 5 in the hundredths place.

Answers (pages 50–51)

Question	Answer	Detailed Explanation
1	60,000 + 8,000 + 900 + 70 + 5	Ten Thousandth TTH TH H T O 6 0 0 0 0 8 0 0 0 9 0 0 7 0 5 6 8 9 7 5
2	A	
3	B, C	$\frac{1}{10}$ of the tenths place Ones Tenth Hundreth Thousandth 0 . 7 0 . 0 7 10 x the tenths place
4	See Detailed Explanation.	3 in the hundreds places is 10 times the value of the 3 in the tens place.
5	See Detailed Explanation.	There are zero thousands. Tyler mistook the 7 to be in the thousands place when it is in the hundreds place.
6	A	Remember "and" represents the decimal when reading decimal numbers.
7	C	8 is in the tenths place and 4 is in the hundredths place. A decimal number is read by the value of the last digit.

Question	Answer	Detailed Explanation
8	10	
9	100	
10	1,000	

Comparing Decimals

You should have a solid foundation in the power of 10 and in determining the value of a digit based on the place value, so comparing decimals should be a cinch.

THE STANDARD

CCSS.5.NBT.3b. Compare two decimals to thousandths based on meanings of the digits in each place, using >, =, and < symbols to record the results of comparisons.

What Does This Mean?

You are able to use your knowledge of the place value system to understand the size of decimal numbers and to compare decimal numbers. You are able to compare tenths to tenths, hundredths to hundredths, thousandths to thousandths in order to determine if one number is greater than, less than, or equal to another decimal number.

What Tool Can Assist You in Comparing Decimal Numbers?

Sometimes it can be difficult to determine which decimal number is the larger of the two. However, using a place value chart will allow you to compare the numbers place by place, tenths to tenths, hundredths to hundredths, thousandths to thousandths until the two numbers differ.

KEY CONCEPT

Compare decimal numbers place by place, tenths to tenths, hundredths to hundredths, and/or thousandths to thousandths until the two numbers differ.

Example 1

Which decimal is greater, 0.52 or 0.519?

HELPFUL HINT

Do not assume a decimal number is greater than another because it has more digits

Hundreds	Tens	Ones	•	Tenths	Hundredths	Thousandths
		0	•	5	2	0
		0	•	5	1	9

In comparing the two decimals, both numbers have a 0 in the ones place and a 5 in the tenths place. However, the 2 in the hundredths place is more than the 1 in the hundredths place, so 0.52 is greater than 0.519.

HELPFUL HINT

Adding a zero to the end of a decimal number does not change the value of the number and allows you to compare place values using the same whole.

Example 2

Which decimal is greater 98.172 or 135.139?

Hundreds	Tens	Ones	●	Tenths	Hundredths	Thousandths
	9	8	●	1	7	2
1	3	5	●	1	3	9

Since both numbers have digits to the left of the decimal, it is not necessary to look at the decimal values. You can see that the second number has a 1 in the hundreds place, making 135.139 greater than 98.172.

Example 3

Which decimal is greater 78.163 or 78.18?

Hundreds	Tens	Ones	●	Tenths	Hundredths	Thousandths
	7	8	●	1	6	3
	7	8	●	1	8	0

In comparing the two decimals, both numbers have a 7 in the tens place, 8 in the ones place, 1 in the tenths place. While there is a 3 in the thousandths place, a zero has been added to the other number just to be able to compare place values. The 3 in the thousandths place is more than the 0 in the other number however, you need to focus on the fact that the 8 in the hundredths place is greater than the 6. In reality, you don't even need to look at the digits in the thousandths place. 78.18 is greater than 78.163 due to the value of the 8 in the hundredths place.

PRACTICE—CHECK UNDERSTANDING: Comparing Decimals

1. Select a symbol to correctly fill the blank of each sentence.

   > = <

   3.845 ________ 3.809

   72.043 ________ 73.034

2. Select the numbers that are greater than 7.612.

   ☐ A. 7.153

   ☐ B. 7.73

   ☐ C. 7.599

   ☐ D. 7.62

   ☐ E. 7.063

   ☐ F. 7.615

3. Select the two answers showing three decimals between 7.5 and 8.5.

   ☐ A. seven and six tenths, 7.9, 8.46

   ☐ B. seven hundredths, 7.067, 8.012

   ☐ C. eight and two tenths, 8.34, 8.487

   ☐ D. seven and four tenths, 7.69, 8.43

4. Write a number that is less than 0.765 in the box below.

   ☐

   Justify your answer: ____________________

   ____

   ____

   ____

   ____

5. John ran 5.375 miles on Monday and 5.38 miles on Wednesday. He told you he ran a longer distance on Monday.

Use your knowledge of comparing decimals and place value to explain why John's statement is incorrect.

____

____

____

____

Answers (page 56)

#	Answer	Detailed Explanation
1	> < 	Numbers are the same until the hundredths place. 4 is greater than 0. Numbers are the same until the ones place. 2 is less than 3.
2	B, D, F	Compare digits by place value to determine which values are greater than 7.612.
3	A, C	Use benchmark decimal numbers to determine numbers between 7.5 and 8.5.
4	0.760, 0.750, 0.701, or any other number less than 0.765	
5	See Detailed Explanation.	Ones, Tenths, Hundredths, Thousandths 5 . 3 7 5 5 . 3 8 The numbers are the same until you get to the hundredths place. The 8 is greater than the 7 so 5.38 is more than 5.375. John ran farther on Wednesday.

Rounding Decimals

If you can compare decimals based on place value, you can surely grasp rounding decimals.

THE STANDARD

CCSS 5.NBT.4. Use place value understanding to round decimals to any place.

What Does This Mean?

In fourth grade, you learned to use place value understanding to round whole numbers to any place.

So, if you were asked to *round 132 to the nearest ten*, would your answer be 100 or 130?

Thousands	Hundreds	Tens	Ones
	1	3	2

See! I think you are getting this numbers thing. When you look at the number 132, you see a 1 in the hundreds column, but you realize you want to round to the *nearest ten*. The digit in the tens column is a 3, and then there is a 2 in the ones column. Stating the nearest ten informs you that you are counting by tens and need to determine if 132 is closer to the number 130 or 140. If you are using place value, then you have to round the number to 130 because the 2 in the ones column lets you know the number is closer to 130. However, let's use the number line below to help you see this even closer. Look at the number line below; is 132 closer in distance to 130 or 140? It is obviously 130.

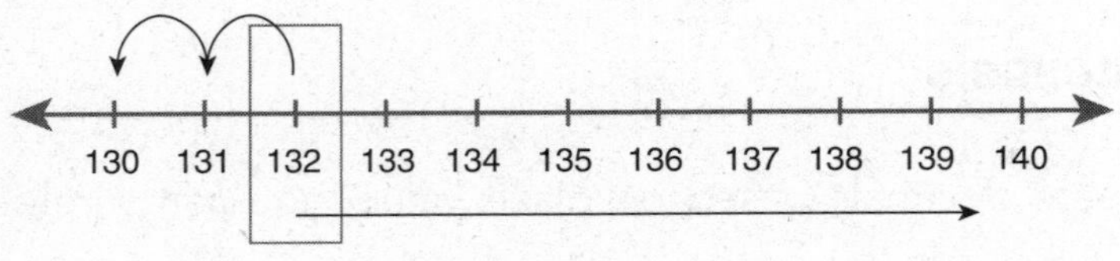

Remember to use benchmark numbers to assist you in rounding. The midway point between 130 and 140 is 135. If the number were greater than 135, it would be considered closer to 140. Since 132 is less than 135 it is closer to 130.

Okay, so what does all this have to do with rounding decimals? While the numbers are going to be less than 1, the process will still be the same. You will now be determining whether to round a decimal number to the nearest tenth, hundredth, or thousandth.

Let's take a shot at rounding a decimal number. Round 0.37 to the nearest tenth. First, let's state the number is read *thirty-seven hundredths*. However, you are rounding to the nearest tenth.

Hundreds	Tens	Ones	•	Tenths	Hundredths	Thousandths
		0	•	3	7	

There is a 3 is in the tenths column and 7 is in the hundredths column. Is the decimal closer to *three tenths* or *four tenths*? Let's use the number line again to get a closer view. Note, the number line is marked off in tenths. It's obviously, 0.4 or four tenths.

What benchmark decimal could you use? ________________

You are correct if you said 0.35. Remember your benchmark numbers.

The answer is obviously 0.4, read as four tenths.

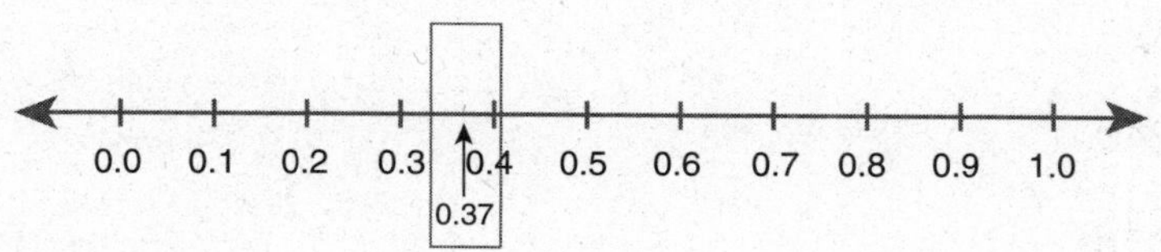

KEY CONCEPT

The answer is 0.4 or four tenths. Drop all other digits, and do not add any zeros. Adding zeros does not change the value of the number, but it changes the way the number is read.

Example

The number 0.37 rounded to the nearest tenth is 0.4. The 4 is in the tenths place. It's important to remember that 0.37 should not become 0.47, which is read forty-seven hundredths. Also, 0.4 should not become 0.40, which is read forty hundredths.

Okay, let's try one more! Round 11.248 to the nearest hundredth.

Hundreds	Tens	Ones	●	Tenths	Hundredths	Thousandths
	1	1	●	2	4	8

1. Which digit below is in the hundredths place?

○ A. 2

○ B. 1

○ C. 4

○ D. 8

2. Which three decimals can be used to help you round the number above?

□ A. 11.250

□ B. 11.235

□ C. 11.240

□ D. 11.245

3. The decimal 11.248 rounded to the nearest hundredth is ________________.

Justify your answer below:

Answers (page 60)

Question	Answer	Detailed Explanation
1	C	4
2	A, C, D	
3	11.25	11.24 11.241 11.242 11.245 11.248 11.25 The digit that needs to be rounded is 4. The 8 in the thousandths place is greater than 5, so I need to round up to the next highest digit in the hundredths place.

PRACTICE—CHECK UNDERSTANDING: Rounding Decimals

1. Round the numbers below to the nearest hundredth.

183.783

12.238

1.207

2. Round the numbers below to the nearest tenth.

5.208

1.671

12.830

3. Use the boxes below. Write four numbers that would round to 13.7.

4. The NASCAR average lap speed is 186.293 miles per hour.

What is the average speed rounded to the nearest tenth?

5. Maximo rounded 0.987 to the nearest tenth and wrote 0.99 on his paper.

Part A. Use your knowledge of rounding decimals and place value to explain why Maximo's statement is incorrect.

____

____

____

____

____

____

Part B. What is the correct answer? Explain how you know.

____

____

____

____

____

____

____

6. The chart below shows the Pole Day time trials of six drivers in the 2013 Indianapolis 500.

Part A: Complete the table.

Driver	Recorded Speed	Speed Rounded to the Nearest Tenth
Driver 12	228.844	
Driver 1	228.282	
Driver 26	228.171	
Driver 3	227.975	
Driver 20	227.952	
Driver 25	227.993	

Part B: The driver's speed determines his starting position on race day. Using the information from the table above, explain whether the recorded speed or rounded speed should be used and why.

____

____

____

____

Answers (pages 61–63)

Question	Answer	Detailed Explanation
1	183.78 12.24 1.21	
2	5.2 1.7 12.8	
3	13.69 13.673 13.712 13.74	
4	186.3	
5	See Detailed Explanation.	A. The number 8 in the hundredths place is greater than 5, so I need to round the tenths digit to the next highest place value. That is the ones place value. Maximo rounded to the nearest hundredth. B. 0.987 rounded to the nearest tenth is 1.0. The decimal 0.987 is closer to one than it is to the 9 in the tenths place.

Question	Answer	Detailed Explanation
6	A. 228.8 228.3 228.2 228.0 228.0 228.0 B. See Detailed Explanation.	B. In racing, the lower qualifying time means you are the fastest car in the field. The recorded speed must be used to determine the starting position. If we used the rounded speed, we would have three drivers qualified for the inside of the first row or pole position for the race.

Multiplying and Dividing Whole Numbers

THE STANDARDS

CCSS.5.NBT.5. Fluently multiply multi-digit whole numbers using the standard algorithm.

CCSS.5.NBT.B.6. Find whole-number quotients of whole numbers with up to four-digit dividends and two-digit divisors, using strategies based on place value, the properties of operations, and/or the relationship between multiplication and division. Illustrate and explain the calculation by using equations, rectangular arrays, and/or area models.

What Does This Mean for Multiplication?

From second grade until now you have been building the knowledge needed to understand what multiplication represents. You discovered that it is really the process of using repeated addition to find your answer. You learned that strategies and practices that involve decomposing numbers (partial products) and using place value, the distributive property, and even area models can assist you in multiplying. Thus far, you have learned the simple way to multiply. However, now, you are going to combine all this understanding to learn the step-by-step process to multiply, called the standard algorithm.

What Does This Mean for Division?

So far you've been using some of the same strategies from the previous page to divide by one-digit divisors. In fifth grade, in addition to the strategies you already know, you are applying place value understanding, the relationship between multiplication and division, and the properties of operations to explain division by two-digit divisors.

What Does "the Simple Way" to Multiply Mean?

You have been multiplying in parts, by multiplying ones and tens separately, in addition to using the properties of operation and addition.

Example

$$275 \times 15 = \_\_\_\_\_\_$$

1. You decomposed the numbers and substituted values that are equivalent to the original numbers:

$$275 = (200 + 70 + 5)$$
$$15 = (10 + 5)$$

2. You used the distributive property to multiply.
3. You used the commutative property of addition. Remember the order of the numbers does not make a difference.

$(200 + 70 + 5) \times (10 + 5) =$

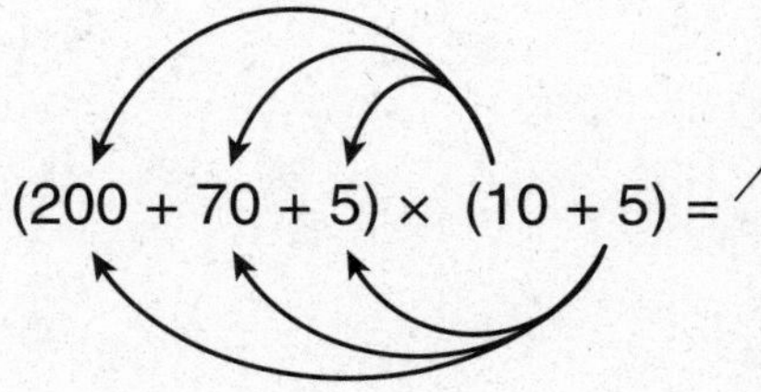

$(200 + 70 + 5) \times (10 + 5) =$

Use the distributive property.

$(200 \times 10) + (70 \times 10) + (5 \times 10) + (200 \times 5) + (70 \times 5) + (5 \times 5) =$

$(2{,}000 + 700 + 50 + 1{,}000 + 350 + 25) =$ use commutative property

$(2{,}000 + 1{,}000 + 700 + 350 + 50 + 25) =$

$(3{,}000 + 1{,}050 + 75) = 4{,}125$

Area Model—Visual

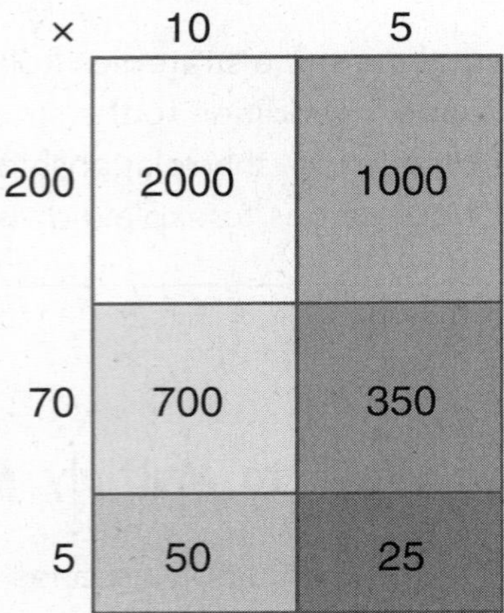

(200×10)+(70×10)+(5×10)+(200×5)+(70×5)+(5×5)=

(2,000 + 700 + 50 + 1000 + 350 + 25) = 4,125

When you use the standard algorithm for multiplication you are decomposing one number, multiplying and adding at the same time. The outcome is the same; it is just a little more compressed.

Example

$$275 \times 15 = \_\_\_\_\_\_\_$$

To multiply in this situation, you would decompose the number 15. So now you have 10 + 5. You should start multiplying 275 by the ones digit of the decomposed number, which would be the 5.

STEP 1: Multiply 5 by the ones digit

$$5 \times 5 = 25$$

$$\begin{array}{r} \scriptsize{2} \\ 27\mathbf{5} \\ \times 1\mathbf{5} \\ \hline \mathbf{5} \end{array}$$

Place 5 in the ones column and write the tens digit, 2, above the tens column. Why? Because you have 25 ones, that can be regrouped as 2 tens and 5 ones.

STEP 2: Multiply 5 by the tens digit, 7, then *add* the 2 tens that you regrouped.

$5 \times 7 = 35 + 2 = 37$

$$\begin{array}{r} {}^{3\,2} \\ 2\mathbf{7}5 \\ \times 1\mathbf{5} \\ \hline \mathbf{7}5 \end{array}$$

HELPFUL HINT

The 7 = 7 tens = 70

The 2 = 2 tens = 20

You are really multiplying:

$5 \times 70 = 350 + 20 = 370$

You are going to write 7 in front of the 5 and write the tens digit, 3, above the hundreds column.

Why? Because you have 37 tens, that can be regrouped as 3 hundreds and 7 tens.

KEY CONCEPT

A common error made when multiplying is multiplying as if each digit represents the value of one. Remember as you move from the ones place to the tens place, and so on, each place value increases times 10.

STEP 3: Multiply 5 by the hundreds digits, 2, then add the 3 hundreds you regrouped.

$5 \times 2 = 10 + 3 = 13$

$$\begin{array}{r} {}^{3\,2} \\ \mathbf{2}75 \\ \times 1\mathbf{5} \\ \hline \mathbf{13}75 \end{array}$$

HELPFUL HINT

The 2 = 2 hundreds = 200

The 3 = 3 hundreds = 300

You are really multiplying and adding as shown below:

$5 \times 200 = 1{,}000 + 300 = 1{,}300$

When you write 13 it represents the 1,300 above.

You are going to write the 13 in front of the 7.

Next you multiply 275 by the digit in the tens place, which would be 1.

HELPFUL HINT

1 ten means we are actually multiplying by 275 × 10.

STEP 4: Multiply 1 ten by the ones digit, 5.

1 × 5 = 5

$$\begin{array}{r} {}^{3\,2} \\ 27\mathbf{5} \\ \times\ \ \mathbf{1}5 \\ \hline 1375 \\ \mathbf{5}0 \end{array}$$

> **HELPFUL HINT**
>
> The 1 = 1 tens = 10
>
> You are really multiplying:
>
> 10 × 5 = 50, so we added a 0 to remind us that we multiplied by the value of ten.

STEP 5: Multiply 1 ten by the tens digit, 7.

1 × 7 = 7

$$\begin{array}{r} {}^{3\,2} \\ 2\mathbf{7}5 \\ \times\ \ \mathbf{1}5 \\ \hline 1375 \\ \mathbf{7}50 \end{array}$$

> **YOUR TURN!**
>
> You are really multiplying:
>
> ☐ × ☐ = ☐

STEP 6: Multiply 1 ten by the hundreds digit, 2.

1 × 2 = 2

$$\begin{array}{r} {}^{3\,2} \\ \mathbf{2}75 \\ \times\ \ \mathbf{1}5 \\ \hline 1375 \\ \mathbf{2}750 \end{array}$$

> **YOUR TURN!**
>
> You are really multiplying:
>
> ☐ × ☐ = ☐

STEP 7: You have finished multiplying so now you need to add the two partial products.

```
   3 2
   275
×   15
------
   1 1
  1375
+2750
------
 4,125
```

Your answer is 4,125.

Add the two partial products remembering to regroup as needed.

Answer

Step 5: 10 × 70 = 700

Step 6: 10 × 200 = 2,000

PRACTICE—CHECK UNDERSTANDING: Multiplying Multi-digit Whole Numbers Using the Standard Algorithm

Find the product using the standard algorithm.

1. 192 × 35	**2.** 623 × 48
3. 705 × 29	**4.** 475 × 53
5. 386 × 53	**6.** 729 × 28
7. 6,273 × 942	**8.** 4,787 × 147

9. 5,297 × 639

10. 2,481 × 573

11. 2,396 × 843

12. 8,147 × 438

Answers (pages 69–70)

Question	Answer	Question	Answer
1	6,720	7	5,909,166
2	29,904	8	703,689
3	20,445	9	3,384,783
4	25,175	10	1,421,613
5	20,458	11	2,019,828
6	20,412	12	3,568,386

What Are the Strategies That Can Be Used to Assist You in Dividing?

Let's look at four strategies that can be used to divide. Many of these strategies require you to utilize your prior understanding of place value, the relationship between multiplication and division, equations, arrays, and the properties of operations.

PARCC will not require you to illustrate division, this is a strategy.

Strategy 1. Using Base-Ten Blocks

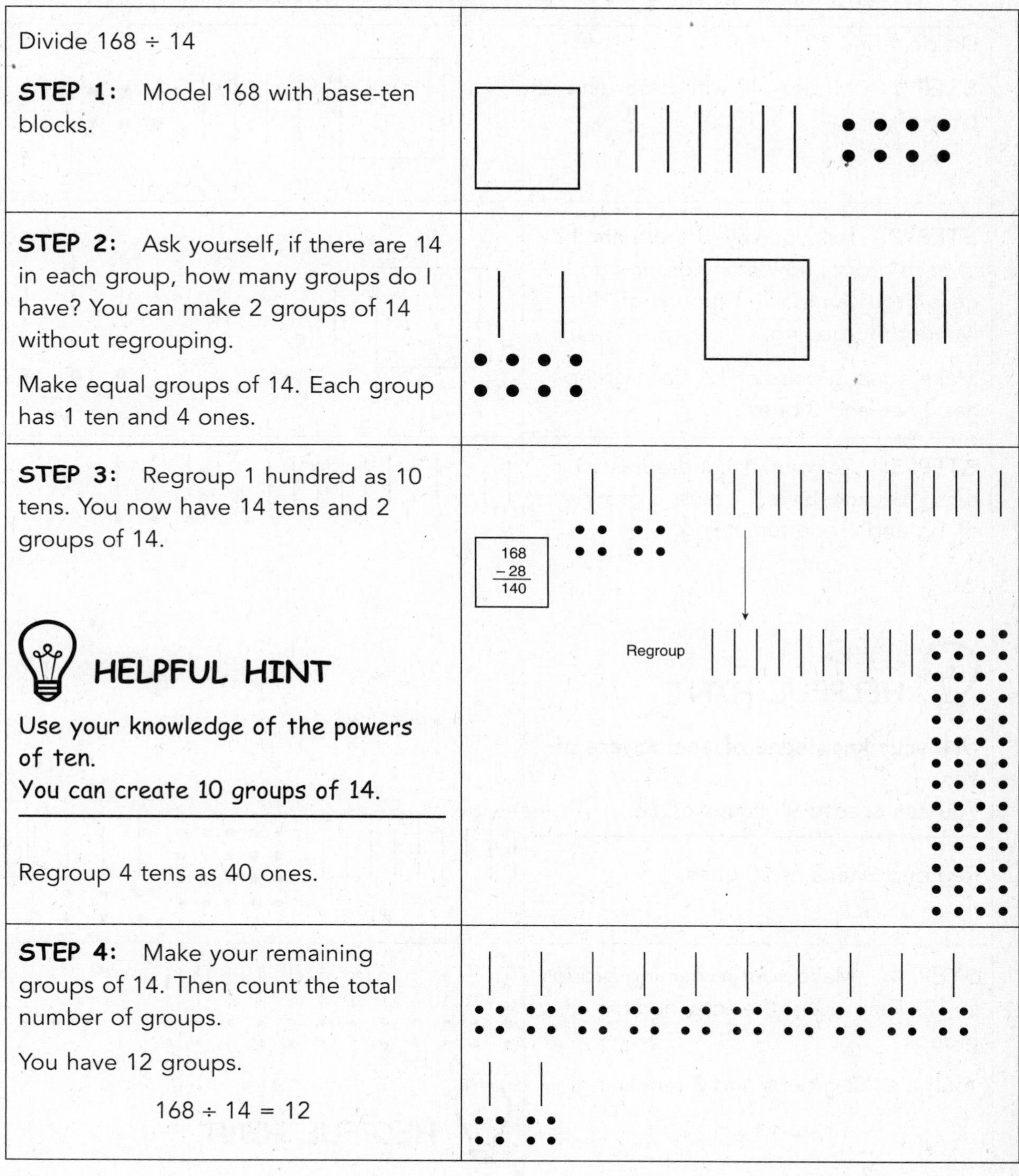

Divide 168 ÷ 14 **STEP 1:** Model 168 with base-ten blocks.	
STEP 2: Ask yourself, if there are 14 in each group, how many groups do I have? You can make 2 groups of 14 without regrouping. Make equal groups of 14. Each group has 1 ten and 4 ones.	
STEP 3: Regroup 1 hundred as 10 tens. You now have 14 tens and 2 groups of 14. **HELPFUL HINT** Use your knowledge of the powers of ten. You can create 10 groups of 14. Regroup 4 tens as 40 ones.	
STEP 4: Make your remaining groups of 14. Then count the total number of groups. You have 12 groups. 168 ÷ 14 = 12	

What Happens When Groups Are Not Evenly Divided?

Divide 147 ÷ 12

STEP 1: Model 147 with base-ten blocks.

STEP 2: Ask yourself, if there are 12 in each group, how many groups do I have? You can make 3 groups of 12 without regrouping.

Make equal groups of 12. Each group has 1 ten and 2 ones.

STEP 3: Regroup 1 hundred as 10 tens. You now have 11 tens, 3 groups of 12, and 1 one remaining.

HELPFUL HINT

Use your knowledge of the powers of ten.
You can create 9 groups of 12.

Regroup 2 tens as 20 ones.

Regroup

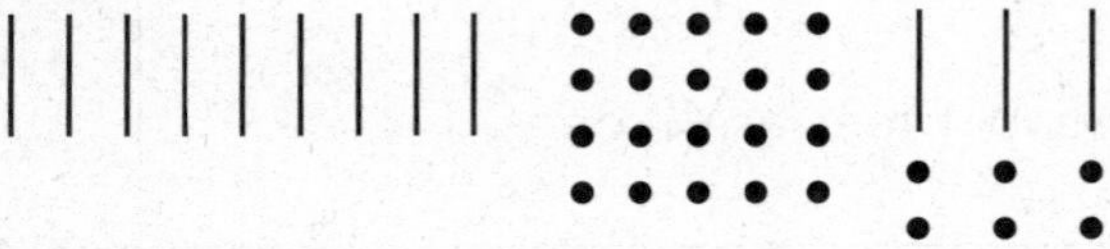

$$\begin{array}{r} 147 \\ -\ 36 \\ \hline 111 \end{array}$$

STEP 4: Make your remaining groups of 12. Then count the total number of groups.

You have 12 groups and 3 remaining.

$$147 \div 12 = 12 \text{ r3}$$

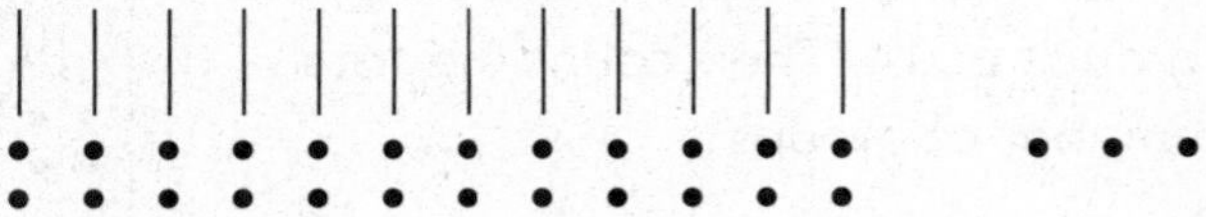

HELPFUL HINT

The remainder is represented by the letter *r*. In the example, r3 means that there are 3 ones left over and that you cannot make another equal group of 12.

PRACTICE—CHECK UNDERSTANDING: Dividing Whole Numbers—Base-Ten Blocks

1. 314 ÷ 12 =

2. 252 ÷ 21 =

3. 768 ÷ 48 =

4. 578 ÷ 32 =

Answers (page 73)

Question	Answer	Detailed Explanation
1	26 r2	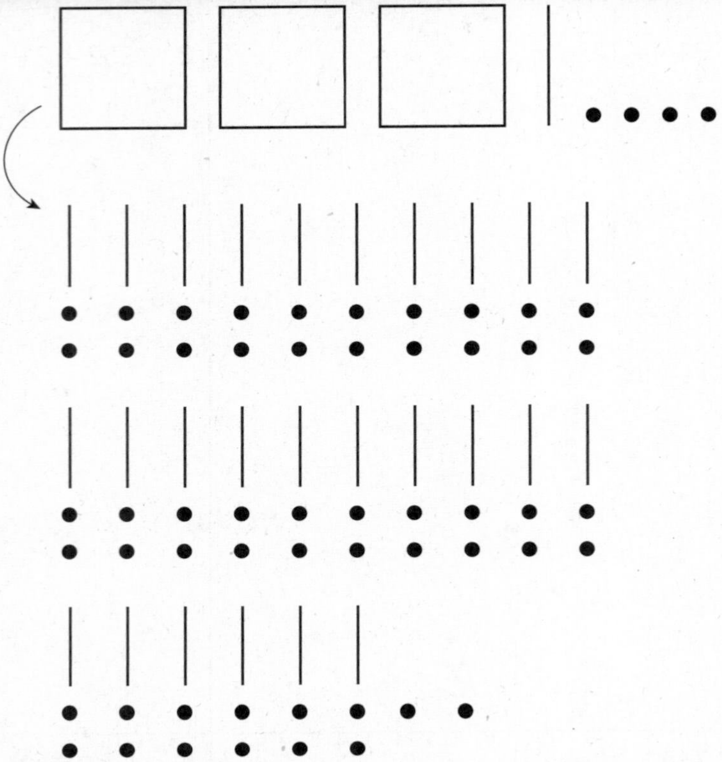
2	12	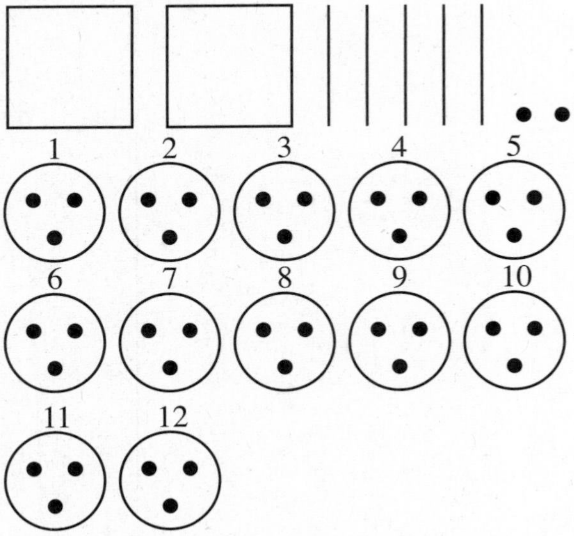 1 2 3 4 5 6 7 8 9 10 11 12

Question	Answer	Detailed Explanation
3	16	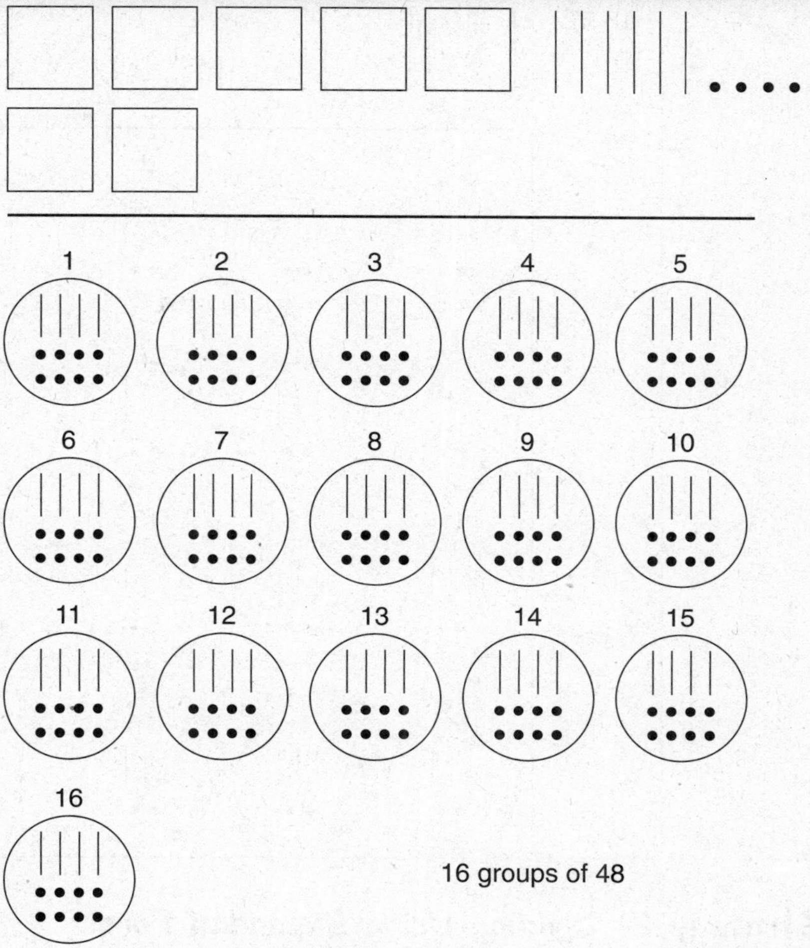 1 2 3 4 5 6 7 8 9 10 11 12 13 14 15 16 16 groups of 48

Question	Answer	Detailed Explanation
4	18 r2	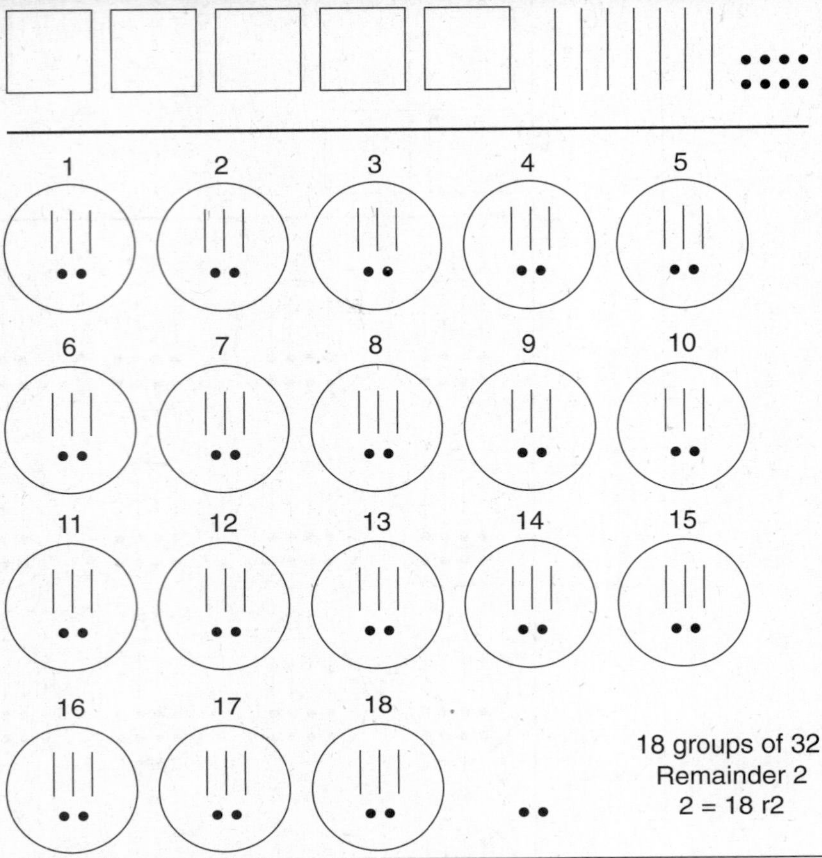 1 2 3 4 5 6 7 8 9 10 11 12 13 14 15 16 17 18 18 groups of 32 Remainder 2 2 = 18 r2

Strategy 2. Dividing Using Expanded Form

Sometimes, breaking a number into smaller parts or expanded form can assist you in dividing. Let's look at the problem below.

Example

$$325 \div 5$$

STEP 1: You can approximate your quotient by rounding the dividend to a number that is lower and higher than your original number, and is easy to divide. In this case the numbers for the dividend would be 300 and 350.

6 × 5 = 30, so **60 × 5 = 300**

7 × 5 = 35, so **70 × 5 = 350**

You know your quotient will be some number between 60 and 70.

STEP 2: Next you can write 325 in expanded form.

(300 + 20 + 5) ÷ 5

STEP 3: Now, you want to handle each number as its own division problem. Let's create a chart that will help to keep your work organized.

Dividend	Divisor	Partial Quotient
300	÷ 5	60
20	÷ 5	4
5	÷ 5	1

STEP 4: Add the partial quotients together to arrive at your answer.

60 + 4 + 1 = 65

STEP 5: Use multiplication to check you work.

✓ 65 × 5 = 325

✓ 325 ÷ 5 = 65

So, 325 can be divided into 65 equal groups.

PRACTICE—CHECK UNDERSTANDING: Dividing Whole Numbers—Expanded Notation

1. 875 ÷ 5 =

2. 1,520 ÷ 20 =

3. 1,320 ÷ 10 =

4. 1,475 ÷ 5 =

Answers (page 77)

<table>
<tr><th>Question</th><th>Answer</th><th>Detailed Explanation</th></tr>
<tr><td>1</td><td>175</td><td><table><tr><th>Dividend</th><th>Divisor</th><th>Partial quotient</th></tr><tr><td>800</td><td>÷ 5</td><td>160</td></tr><tr><td>70</td><td>÷ 5</td><td>14</td></tr><tr><td>5</td><td>÷ 5</td><td>1</td></tr><tr><td></td><td></td><td>175</td></tr></table></td></tr>
<tr><td>2</td><td>76</td><td><table><tr><th>Dividend</th><th>Divisor</th><th>Partial quotient</th></tr><tr><td>1000</td><td>÷ 20</td><td>50</td></tr><tr><td>500</td><td>÷ 20</td><td>25</td></tr><tr><td>20</td><td>÷ 20</td><td>1</td></tr><tr><td></td><td></td><td>76</td></tr></table></td></tr>
<tr><td>3</td><td>132</td><td><table><tr><th>Dividend</th><th>Divisor</th><th>Partial quotient</th></tr><tr><td>1000</td><td>÷ 10</td><td>100</td></tr><tr><td>300</td><td>÷ 10</td><td>30</td></tr><tr><td>20</td><td>÷ 10</td><td>2</td></tr></table></td></tr>
<tr><td>4</td><td>295</td><td><table><tr><th>Dividend</th><th>Divisor</th><th>Partial quotient</th></tr><tr><td>1000</td><td>÷ 5</td><td>200</td></tr><tr><td>400</td><td>÷ 5</td><td>80</td></tr><tr><td>70</td><td>÷ 5</td><td>14</td></tr><tr><td>5</td><td>÷ 5</td><td>1</td></tr><tr><td></td><td></td><td>295</td></tr></table></td></tr>
</table>

Strategy 3. Using an Area Model

Again we are going to look at breaking numbers into smaller parts, however, this time using an area model.

HELPFUL HINT

You can approximate your quotient by rounding the dividend to a number that is close to your original number, and is easy to divide.

455 ÷ 15

15 × 3 = 45, so 15 × 30 = 450

You know your quotient will be a little more than 30.

Divide 455 ÷ 15

STEP 1: You know 15 × 2 = 30

15 × 20 = 300. Once you subtract, you now have 155 remaining.

	15
20	300

15 × 20 = 300

$$\begin{array}{r} 455 \\ -\,300 \\ \hline 155 \end{array}$$

STEP 2: You know 15 × 10 = 150. Once you subtract, you see there are no remaining equal groups of 15, you have 5 remaining.

	15
20	300
10	150

15 × 10 = 150

$$\begin{array}{r} 455 \\ -\,300 \\ \hline 155 \\ -\,150 \\ \hline 5 \end{array}$$

Remaining

STEP 3: Add your partial quotients to get your answer:

20 + 10 = 30 and the 5 remaining. 30 r5

30 × 15 = 450 + 5 = 455

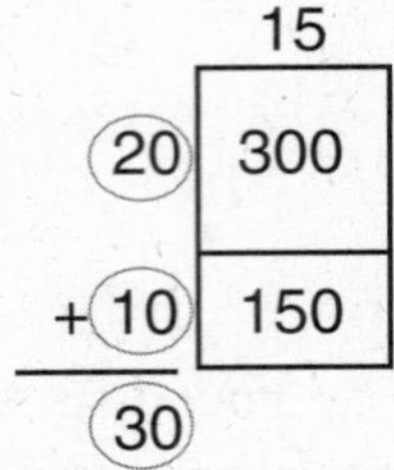

455
− 300
155
− 150
5
Remaining

20 + 10 = 30 r5

PRACTICE—CHECK UNDERSTANDING: Dividing Whole Numbers—Area Model

1. 1,365 ÷ 12 =

2. 2,784 ÷ 22 =

3. 552 ÷ 46 =

4. 875 ÷ 35 =

Answers (page 80)

Question	Answer	Detailed Explanation
1	113 r9	12 (100) 1200 (12 × 100) (10) 120 (12 × 10) +(3) 36 (12 × 3) (113) 1365 − 1200 = 165 165 − 120 = 45 45 − 36 = 9 left
2	126 r12	12 (100) 2200 22 × 100 = 2200 (20) 440 22 × 20 = 440 +(6) 132 22 × 6 = 132 (126) 2784 − 2200 = 584 584 − 440 = 144 144 − 132 = 12 left
3	12	46 (10) 460 46 × 10 = 460 +(2) 92 46 × 2 = 92 (12) 562 − 460 = 92 92 − 92 = 0
4	25	35 (20) 700 35 × 20 = 700 +(5) 175 35 × 5 = 175 (25) 875 − 700 = 175 175 − 175 = 0

Strategy 4. Using the Relationship Between Multiplication and Division/Using the Distributive Property

492 ÷ 4

STEP 1: Use inverse operations to write the related multiplication problem for the division problem.

4 × _____ = 492

STEP 2: Use the distributive property to decompose the number into numbers that are multiples of the divisor.

HELPFUL HINT

Always use a multiple of 10 for one of the numbers.

400 + 92 = 492

(4 × **100**) + (4 × **23**) = 492

4 × (**100** + **23**) = 492

STEP 3: Add the numbers in parentheses to find the sum of the unknown factor.

(**100** + **23**) = 123

STEP 4: Write the division sentence with your solution.

492 ÷ 4 = 123

PRACTICE—CHECK UNDERSTANDING: Dividing Whole Numbers—Relationship of Multiplication to Division

1. $105 \div 7 =$

2. $2{,}025 \div 15 =$

3. $636 \div 3 =$

4. $875 \div 35 =$

Answers (page 83)

Question	Answer	Detailed Explanation
1	15	7 × ____________ = 105 70 + 35 (7 × **10**) + (7 × **5**) 7 × (**10 + 5**) 10 + 5 = 15 105 ÷ 7 = 15
2	135	15 × ____________________ = 2,025 1,500 + 525 (15 × **100**) + (15 × **35**) 15 × (**100 + 35**) 100 + 35 = 135 2,025 ÷ 15 = 135
3	212	3 × _______ = 636 600 + 36 (3 × **200**) + (3 × **12**) 3 × (**200 + 12**) 200 + 12 = 212 636 ÷ 3 = 212
4	25	35 × ___________ = 875 700 + 175 (35 × **20**) + (35 × **5**) 35 × (**20 + 5**) 20 + 5 = 25 875 ÷ 35 = 25

Adding and Subtracting Decimals

THE STANDARD

CCSS.5.NBT.7. Add, subtract, multiply, and divide decimals to hundredths, using concrete models or drawings and strategies based on place value, properties of operations, and/or the relationship between addition and subtraction; relate the strategy to a written method and explain the reasoning used.

What Does That Mean?

You will be able to apply visual representations and the strategies of place value in addition to the properties of operations used with whole numbers to add, subtract, multiply, and divide decimals and be able to explain why.

What Models Help You Add and Subtract Decimals?

You can model decimal addition and subtraction by using base-ten blocks as well as a place value chart. Below you have a visual representation of 1, 0.1, and 0.01.

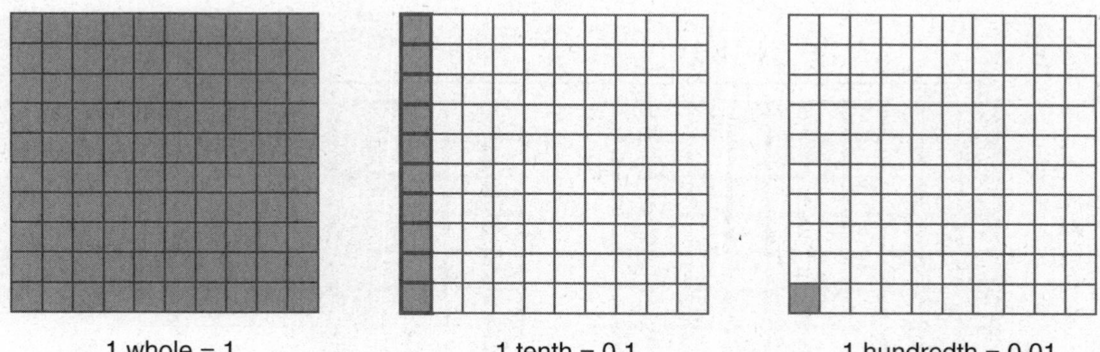

1 whole = 1 1 tenth = 0.1 1 hundredth = 0.01

You can use the models above to help you add and subtract decimals.

Addition

Example 1

Add 1.4 + 0.22

1.4 + 0.22 =

STEP 1: Shade squares to represent the number 1.4.

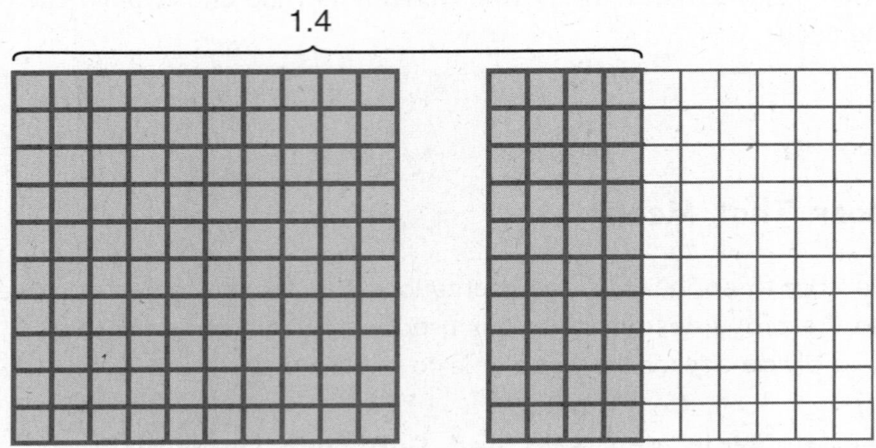

STEP 2: Shade additional squares to show adding 0.22.

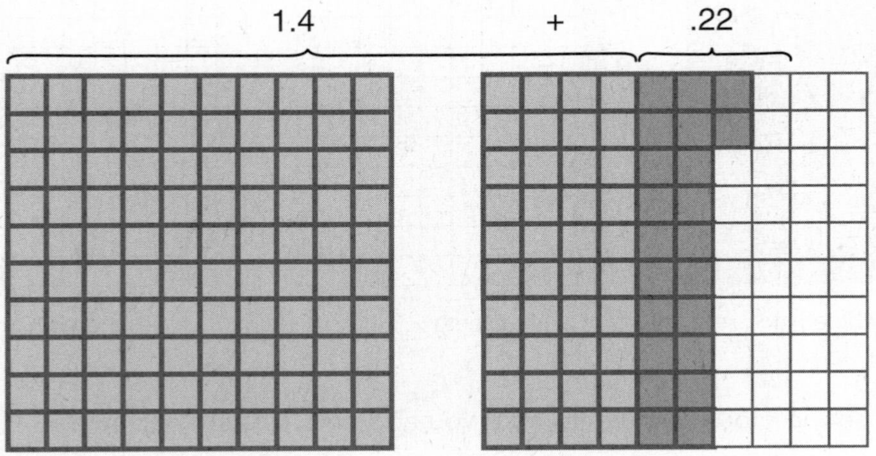

STEP 3: Count the total number of squares. You have 1 whole square and 62 one-hundredths squares shaded

1.4 + 0.22 = 1.62

Example 2

Line up the decimals to add 1.4 + 0.22.

STEP 1: Line up the place values for each number in the place value chart.

		Ones	•	Tenths	Hundredths	
		1	•	4	0	
	+	0	•	2	2	

HELPFUL HINT

Remember adding a zero as a place holder does not change the value of the decimal.

STEP 2: Then add the number as normal. Make sure you have a decimal point in your answer.

		Ones	•	Tenths	Hundredths	
		1	•	4	0	
	+	0	•	2	2	
		1	•	6	2	

Subtraction

Example 1

Subtract 1.7 – 0.64.

1.7 – 0.64

STEP 1: Shade squares to represent the number 1.7.

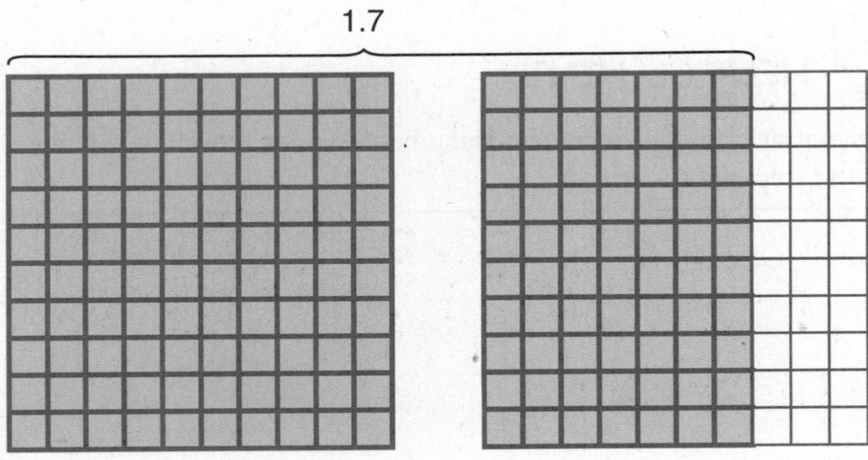

STEP 2: Outline 64 of the shaded squares and put an X over them to represent subtraction.

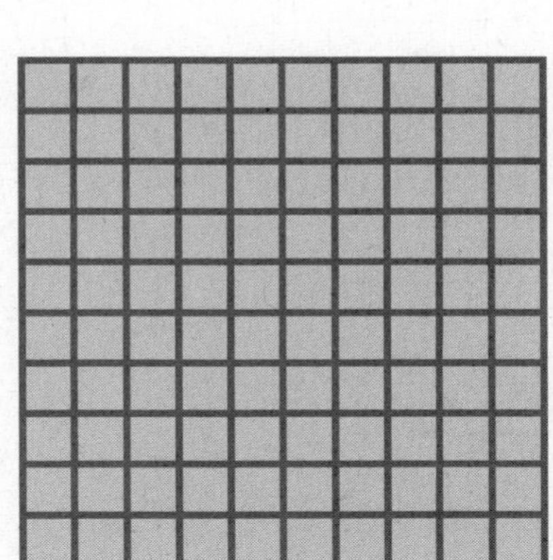

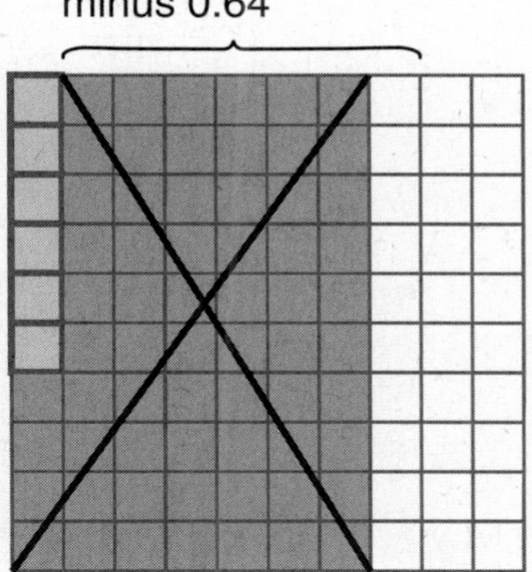

HELPFUL HINT

Outlining and putting an X through the squares allows you to see how many squares have been subtracted.

STEP 3: Count the remaining shaded squares.

1.7 – 0.64 = 1.06

Example 2

Line up the decimals to subtract 1.7 – 0.64.

STEP 1: Line up the place values for each number in the place value chart.

HELPFUL HINT

You will need to add a zero in the hundredths place as a place holder.

		Ones	•	Tenths	Hundredths	
		1	•	7	0	
	–	0	•	6	4	

STEP 2: Regroup 1 tenth for 10 hundredths and subtract. Make sure you have a decimal point in your answer.

		Ones	•	Tenths	Hundredths	
		1	•	7	0	
	–	0	•	6	4	

		Ones	•	Tenths	Hundredths	
		1	•	6	10	
	–	0	•	6	4	
		1	•	0	6	

PRACTICE—CHECK UNDERSTANDING: Add or Subtract the Decimals

Find the sum or difference. Use base-ten blocks or a place value chart if needed. Base-ten block sheets and place value charts are located in Appendix B.

1.
$$\begin{array}{r} 1.76 \\ +.35 \\ \hline \end{array}$$

2. 6.23 – 1.48

3. 15.05 – 2.9

4.
$$\begin{array}{r} 43.75 \\ -28.53 \\ \hline \end{array}$$

5.
$$\begin{array}{r} 0.86 \\ +0.59 \\ \hline \end{array}$$

6.
$$\begin{array}{r} 27.29 \\ +43.28 \\ \hline \end{array}$$

7. $\begin{array}{r} 62.73 \\ -9.42 \\ \hline \end{array}$

8. 165.87 + 14.7

9. $\begin{array}{r} 529.7 \\ + \quad 6.39 \\ \hline \end{array}$

10. $\begin{array}{r} 64.81 \\ +9.03 \\ \hline \end{array}$

11. $\begin{array}{r} 23.96 \\ +8.43 \\ \hline \end{array}$

12. 81.47 – 43.88

Answers (pages 91–92)

Question	Answer
1	2.11
2	4.75
3	12.15
4	15.22
5	1.45
6	70.57
7	53.31
8	180.57
9	536.09
10	73.84
11	32.39
12	37.59

What Method Can You Use to Determine the Reasonableness of Your Answer?

Rounding decimals to the nearest whole number is a great way to estimate the sums and differences.

Example 1

Round to the nearest whole number and add.

$$\begin{array}{r} 23.96 \\ +8.43 \\ \hline \end{array} \rightarrow \begin{array}{r} 24 \\ +8 \\ \hline 32 \end{array} \rightarrow \boxed{\text{The sum is about 32.}}$$

Example 2

Round to the nearest whole number and subtract.

$$\begin{array}{r} 5.76 \\ -0.83 \\ \hline \end{array} \rightarrow \begin{array}{r} 6 \\ -1 \\ \hline 5 \end{array} \rightarrow \boxed{\text{The difference is about 5.}}$$

PRACTICE—CHECK UNDERSTANDING: Adding and Subtracting Decimals—Estimating the sum or difference.

1. 7.26 + 1.75

2. 9.53 – 1.74

3. 15.05 – 12.9

4.
$$\begin{array}{r} 43.15 \\ -28.33 \\ \hline \end{array}$$

Answers (page 94)

Question	Answer	Detailed Explanation
1	9	7.26 → 7 1.75 → 2
2	8	10 – 2
3	2	15 – 13
4	15	43 – 28

Multiplying and Dividing Decimals

What Models Help You Multiply and Divide Decimals?

You can model decimal multiplication and division by using base-ten blocks.

Multiplication

Example

Multiply 0.9 × 0.5.

STEP 1: Shade 9 rows to represent 0.9.

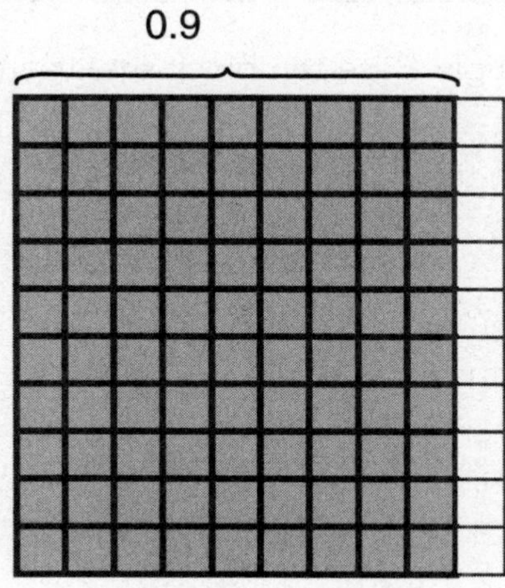

STEP 2: Shade 5 columns to represent 0.5.

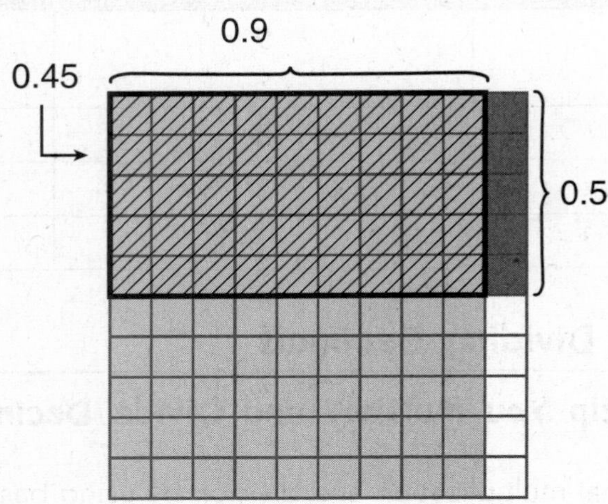

STEP 3: Count the number of squares that overlap.

$$0.9 \times 0.5 = 0.45$$

PRACTICE—CHECK UNDERSTANDING: Multiplying Decimals

Use base-ten blocks to multiply. Base-ten block sheets are provided in Appendix B.

1. $0.3 \times 0.3 =$

2. $0.5 \times 0.7 =$

3. $2.5 \times 0.8 =$

4. $\begin{array}{r} 1.4 \\ \times 0.7 \\ \hline \end{array}$

Answers (page 96)

Question	Answer	Detailed Explanation
1	0.09	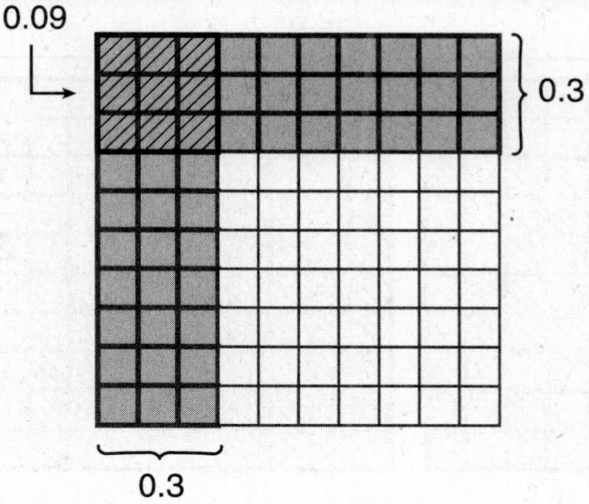 0.09; 0.3; 0.3
2	0.35	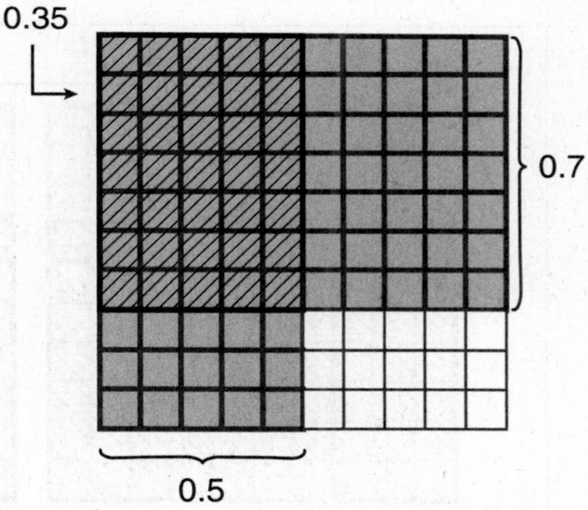 0.35; 0.7; 0.5

Question	Answer	Detailed Explanation
3	2	
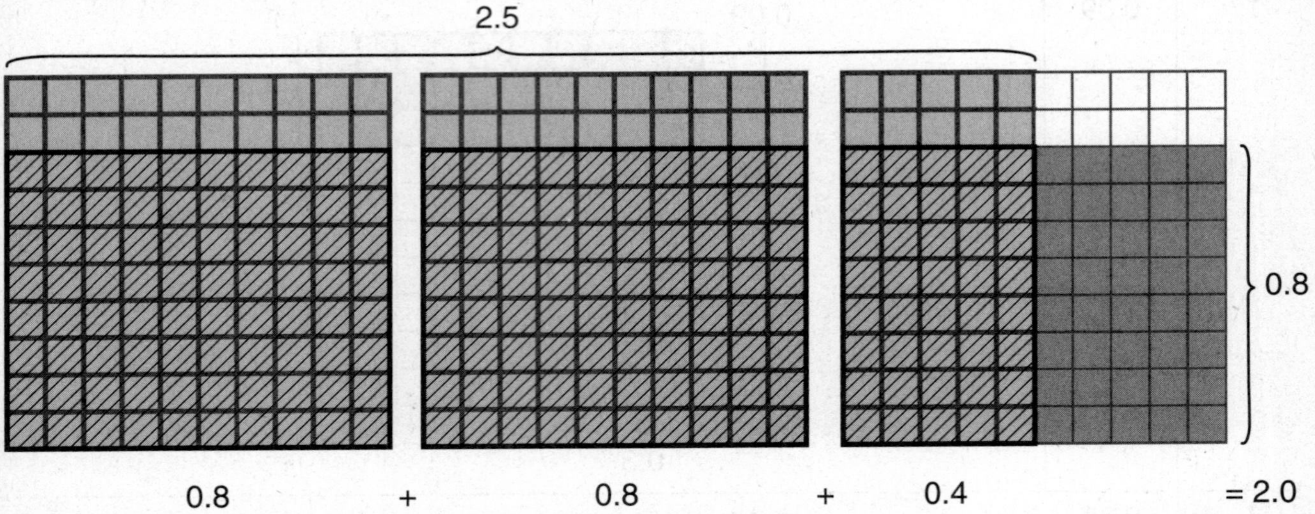 2.5 0.8 + 0.8 + 0.4 = 2.0 0.8		
4	.98	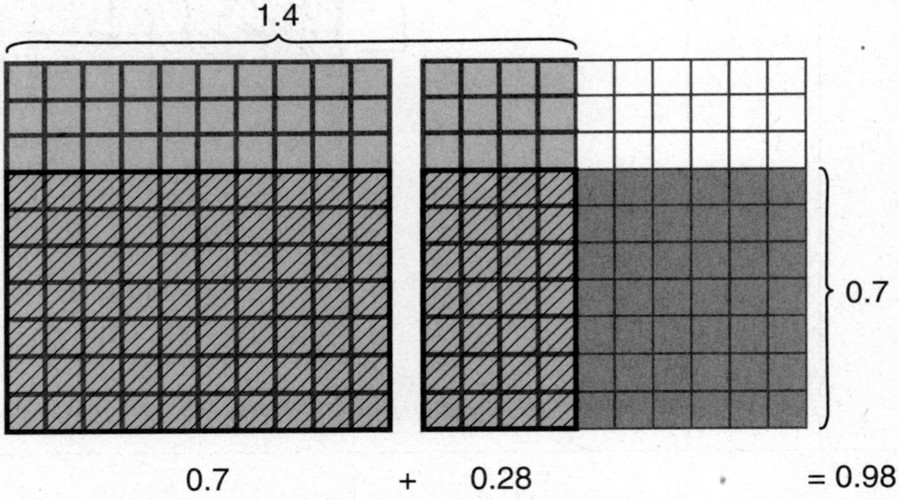 1.4 0.7 0.7 + 0.28 = 0.98

Placing the Decimal Point

Example

Multiply 5.3 × 2.71.

STEP 1: Multiply as if you are multiplying whole numbers.

$$\begin{array}{rl} 271 & \\ \times 53 & \\ \hline 813 & (271 \times 3)\text{ Partial Product} \\ +13550 & (271 \times 50)\text{ Partial Product} \\ \hline 14363 & \end{array}$$

STEP 2: Add the number of decimal places in the factors to determine where to place the decimal point in the product.

$$\begin{array}{rl} 2.71 & \\ \times 5.3 & \leftarrow \boxed{\begin{array}{l} 2\text{ decimal places} \\ +1\text{ decimal place} \end{array}} \\ \hline 813 & \\ +13550 & \\ \hline 14.363 & \leftarrow \boxed{3\text{ decimal places}} \end{array}$$

So 5.3 × 2.71 = 14.363.

PRACTICE—CHECK UNDERSTANDING: Multiplying Decimals—Placing the Decimal Point

1. $\begin{array}{r} 12.3 \\ \times 0.5 \\ \hline \end{array}$

2. $\begin{array}{r} 0.05 \\ \times 0.7 \\ \hline \end{array}$

3.

$$\begin{array}{r} 4.5 \\ \times\ 2.8 \\ \hline \end{array}$$

4.

$$\begin{array}{r} 1.4 \\ \times 0.8 \\ \hline \end{array}$$

5.

$$\begin{array}{r} 2.4 \\ \times 0.6 \\ \hline 144 \end{array}$$

6.

$$\begin{array}{r} 3.42 \\ \times\ 0.7 \\ \hline 2394 \end{array}$$

7.

$$\begin{array}{r} 1.04 \\ \times\ 0.8 \\ \hline 0832 \end{array}$$

8.

$$\begin{array}{r} 15.4 \\ \times\ 0.8 \\ \hline 1232 \end{array}$$

Answers (pages 99–100)

#	Answer
1	6.15
2	0.035
3	12.6
4	1.12
5	1.44
6	2.394
7	0.832
8	12.32

Division

Example 1

Divide 1.2 ÷ 0.4.

STEP 1: Shade 12 tenths. This represents the dividend 1.2.

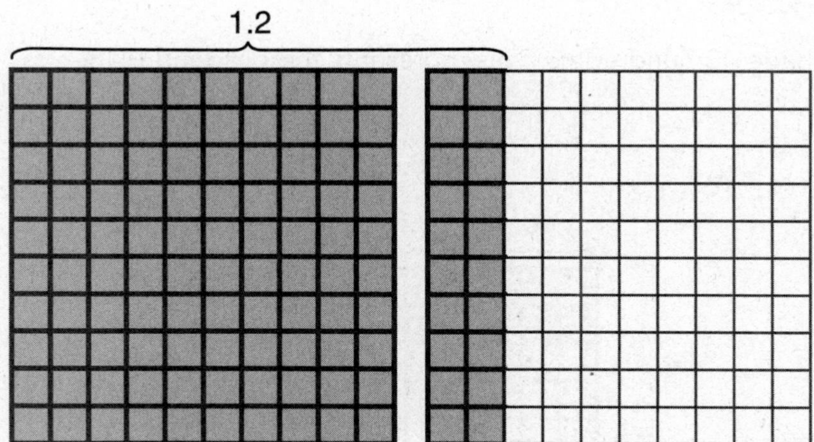

STEP 2: Divide the 12 tenths into groups of 4 tenths. In this example, you are dividing by 0.4 or taking the 12 tenths and creating equal groups of 4 tenths.

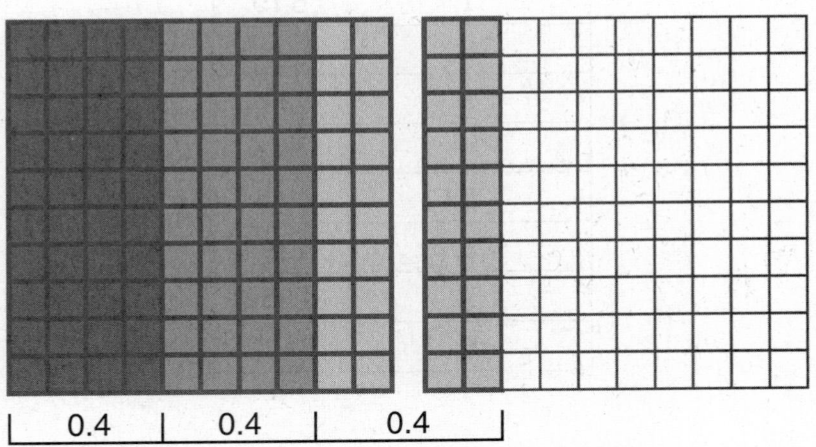

STEP 3: Count the groups of 4 tenths in 12 tenths.

12 tenths ÷ 4 tenths = 3

Example 2

Divide 0.48 ÷ 0.06.

STEP 1: Shade 48 hundredths. This represents the dividend 0.48.

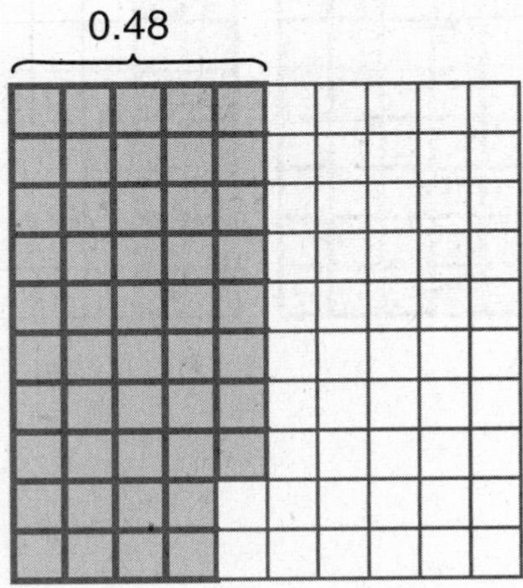

STEP 2: Divide the 48 hundredths into groups of 6 hundredths. This represents your divisor of 0.06.

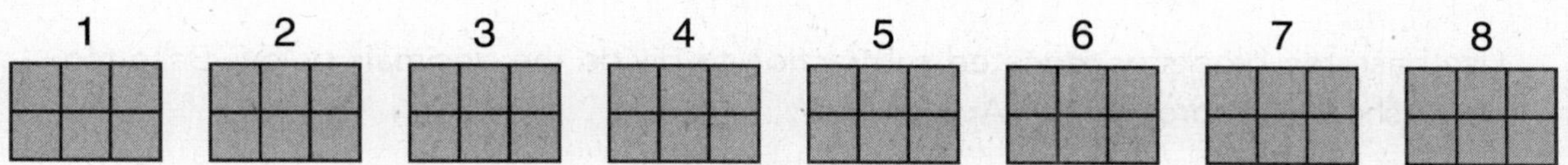

STEP 3: Count the groups of 6 hundredths in 48 hundredths.

48 hundredths ÷ 6 hundredths = 8

Example 3

Divide 2.4 ÷ 0.3.

24 tenths

3 tenths	3 tenths	3 tenths	3 tenths	3 tenths	3 tenths	3 tenths	3 tenths

8 groups

STEP 1: Use repeated subtraction to find the number of groups.

1 group of 0.3 — 2.4 – 0.3 = 2.1
2 groups — 2.1 – 0.3 = 1.8
3 groups — 1.8 – 0.3 = 1.5
4 groups — 1.5 – 0.3 = 1.2
5 groups — 1.2 – 0.3 = 0.9
6 groups — 0.9 – 0.3 = 0.6
7 groups — 0.6 – 0.3 = 0.3
8 groups — 0.3 – 0.3 = 0

PRACTICE—CHECK UNDERSTANDING: Dividing Decimals

Use base-ten blocks or repeated subtraction to divide the decimals below. Base-ten block sheets are provided in Appendix B.

1. 6.4 ÷ 0.4

2. 0.28 ÷ 0.07

3. 0.84 ÷ 0.14

4. 8.4 ÷ 2.1

Answers (page 104)

Question	Answer	Detailed Explanation
1	16	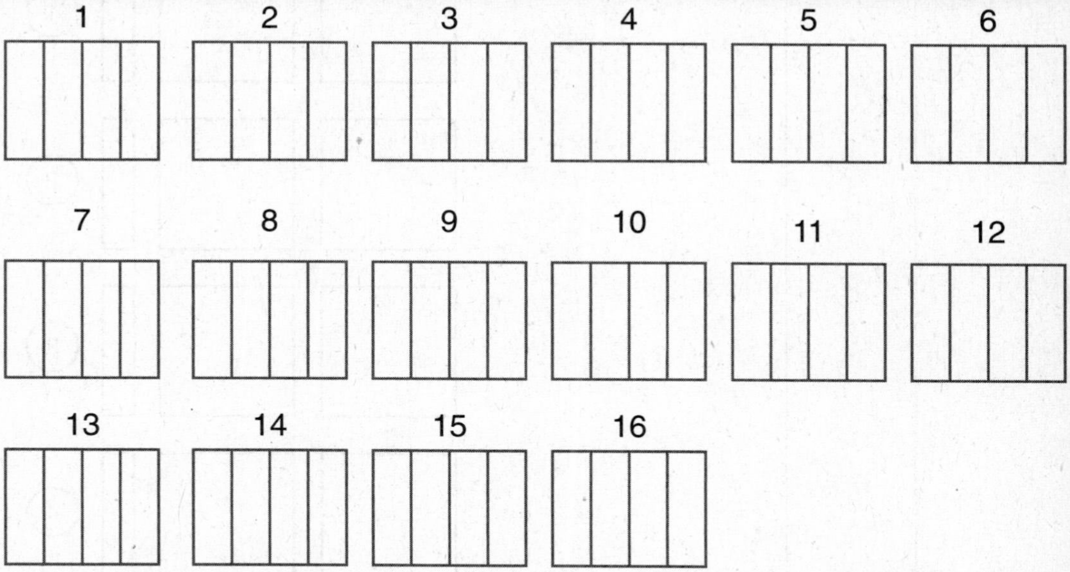 1 2 3 4 5 6 7 8 9 10 11 12 13 14 15 16
2	4	 1 2 3 4
3	6	Count the groups of 14 hundredths 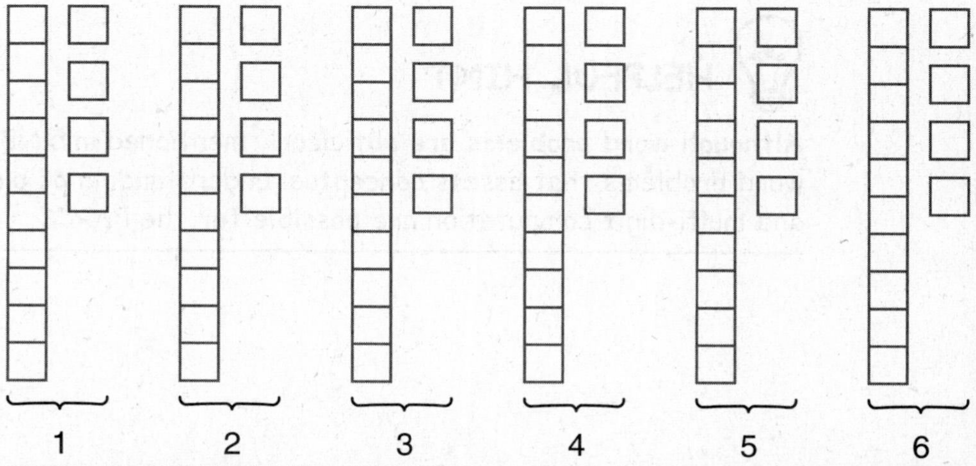 1 2 3 4 5 6

Question	Answer	Detailed Explanation
4	4	① ② ③ ④

Problem-Solving 5.NBT

PRACTICE—APPLICATION: Numbers Base-Ten

HELPFUL HINT

Although word problems are not clearly mentioned in 5.NBT, simple word problems that assess conceptual understanding of place value and multi-digit computation are possible for the PARCC.

1. There were 1,275 students registered for the robotics club competition. Each student paid a $103 entrance fee. How much money was paid in all?

2. A local car wash reported that 1,560 cars entered the car wash over a 12-day period. How many cars were washed per day?

3. The track team ran 3.75 miles in the morning and 2.4 miles in the afternoon. How many miles did they run altogether?

4. Felix was 49.5 inches tall last year. This year he is 54.25 inches tall. How much taller did he grow in a year?

5. Gas costs \$2.35 a gallon. You need 1.4 gallons of gas for your riding lawn mower. How much do you pay for gas?

6. Kayla is making cloth book covers for the school fair. She has 10.4 yards of material. Each cover uses 0.4 yards of material. How many book covers can she make?

7. Michelle looks at the number sentence below and determines that 5.405 is greater than 5.408.

5.405 > 5.408

- Explain why Michelle's answer is not reasonable.
- What should the correct answer be?
- Explain how to use place value understanding to show your answer is correct.

Answers (pages 107–108)

1. $131,325
2. 130 cars
3. 6.15 miles
4. 4.75 inches taller
5. $3.29 for gas
6. 26 book covers
7.
   - Michelle's answer is not reasonable because she did not compare the value of the digits in the thousandths place correctly. The numbers are the same until the thousandths place, 8 > 5 so her answer is incorrect.
   - The correct answer should be:

     5.405 < 5.408

   - I used a place value chart and compared the digits for each number.

Hundreds	Tens	Ones	•	Tenths	Hundredths	Thousandths
		5	•	4	0	5
		5	•	4	0	8

Operations and Algebraic Thinking

CHAPTER 4

What is the best way to think about this chapter, Operations and Algebraic Thinking? Think in terms of being a doctor.

When you go to the doctor, if you are coughing, the doctor evaluates your symptoms and determines what strategies he will use to solve your problem. While no surgery may have to be performed, the doctor still performed an operation to arrive at the solution. He may have used the stethoscope to listen to your chest as you coughed. He may have asked you questions about whether you cough more during the day, at night, when you are outside, and so on. The major point is that after the doctor has done a lot of thinking about your symptoms he is left with no choice other than to perform an operation if he is going to solve your problem.

Consider yourself a Math Doctor for this chapter. You will be presented with problems that require you to identify strategies that you can use to perform operations to solve problems. You will need to do a lot of thinking about the problems and evaluate which strategies will best help you arrive at a solution.

Start performing those operations, Math Doctor!

Evaluating and Writing Expressions Using Grouping Symbols

THE STANDARDS

CCSS 5.OA.1. Use parentheses, brackets, or braces in numerical expressions, and evaluate expressions with these symbols.

CCSS 5.OA.2. Write simple expressions that record calculations with numbers, and interpret numerical expressions without evaluating them. For example, express the calculation "add 8 and 7, then multiply by 2" as 2 × (8 + 7). Recognize that 3 × (18,932 + 921) is three times as large as 18,932 + 921, without having to calculate the indicated sum or product.

What Does This Mean?

In third grade, you learned the meaning of numerical expressions and the operations symbols through using pictures, objects, words, numbers, and equations. For example, the mathematical equation $3 \times 6 = 6 \times 3$ can be represented as a picture like the one below.

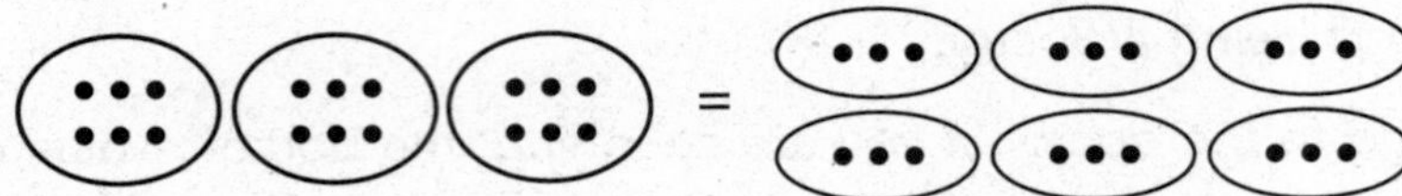

In other words, you know that 3 groups of 6 objects is equivalent to 6 groups of 3 objects because in third grade you learned to identify the meaning of multiplication and equivalence through drawing pictures, counting objects, and so forth. You will expand on your understanding of the language of mathematics. You will learn that the grouping of mathematical operation symbols, numbers, words and parentheses, brackets, or braces translate into some value.

What Do We Mean When We Say Write or Evaluate a Numerical Expression Using Parentheses, Brackets, or Braces?

A **numerical expression** represents a single value. It consists of one or more numbers and operations. Sometimes the numerical expression is disguised as a word problem. At times you have to use **grouping symbols** to identify the order in which you should approach **solving (evaluating)** the problem.

Grouping Symbols

1. Parentheses ( ) Most commonly used
2. Brackets [ ] Used when there is more than one grouping in a problem
3. Braces { } Used when there is more than one grouping in a problem

Written Numerical Expression	Word Problem
(38 − 3) ÷ 5	Samantha has 38 pieces of candy. You have 3 pieces less than Samantha and divide yours equally into 5 groups to make sure you have a treat after school each day this week. How many pieces of candy are in each group?

HELPFUL HINT

Many times your teachers try to assist you by teaching you that there are *key words* that mean certain math terms. While sometimes this may help, you should pay close attention to the order of the words in a sentence to understand the true meaning of the math sentence.

For example in the previous word problem, "three pieces less than Samantha" represents the math sentence 38 – 3. Additionally, you need to use grouping symbols to determine how many pieces of candy you have before you can actually divide the candy into equal groups.

KEY CONCEPT

Grouping symbols identify which operation to do first. However, do not forget that sometimes you will still need to follow the *order of operations*, too.

PRACTICE—CHECK UNDERSTANDING: Evaluate or Write Simple Expressions

Evaluate or write the expressions.

1. $3 \times (6 + 7)$

2. $25 \times [3 + (7 \times 4)] - 18$

3. $24 \div [(2 \times 5) - 2]$

4. $(12 \times 3) + 7 - 11$

5. $4 \times (25 - 20 \div 4)$

6. $3 \times [8 + (48 \div 6 + 3)]$

7. $24 \div [(2 \times 5 - 2) - 2]$

8. $(12 \times 3 \div 9) + 7 - 11$

9. You have a 3-shelf bookcase in your room. 5 books are on each shelf, and 4 are sitting on the floor. Write the expression to find the total number of books you have in your room.

10. You baked 24 cupcakes for each fifth grade class. There are 6 classes. Your brother and his friend ate 5 cupcakes. Write an expression to find the total number of cupcakes that remain.

11. "The product of 7 and 12 divided by 4"

12. "17 less than the quotient of 54 divided by 2"

13. "The sum of 210 and 14 multiplied by 3"

14. "Divide the product of 8 and 9 by the sum of 1 and 3"

15. Alejandro evaluated the expression below and arrived at an answer of 11.

$$2 \times (8 - 5)$$

Use your knowledge of evaluating expressions with grouping symbols to explain why Alejandro's work is incorrect.

____

____

____

____

____

____

____

16. Write the expression that matches each statement in the correct box. Each may be used more than once or not at all.

$(12 \times 4) - 9$	$9 \times (12 - 4)$	$(12 - 9) \times 4$

"Four less than 12 multiplied by 9"	"Nine less than the product of 12 and 4"	"The difference between 12 and 9 multiplied by 4"

Answers (pages 113–115)

Question	Answer	Detailed Explanation
1	39	You should add the numbers in parentheses before you multiply by 3.
2	757	When there are two levels of grouping symbols, complete the parentheses first and then the brackets. Step 1: 25 × [3 + 28] – 18 multiply 7 × 4 Step 2: 25 × [31] – 18 add 28 + 3 Step 3: 775 – 18 multiply 25 × 31 Answer: 757 subtract 18
3	3	Following steps from #2 the problem represents 24 ÷ 8.
4	32	36 + 7 – 11 = 43 – 11 = 32
5	80	Be careful! Even though there is only one set of grouping symbols, you have to make sure you follow the order of operations. Step 1: 4 × (25 – 5) Divide 20 by 4 Step 2: 4 × (20) 25 minus 20 Answer: 80 multiply 20 × 4
6	57	Follow order of operations and system of using grouping symbols.
7	4	Follow order of operations and system of using grouping symbols.
8	0	Following order of operations 12 × 3 = 36 Divide 36 by 9 = 4 Add 4 + 7 = 11 Subtract 11 – 11 = 0
9	(3 × 5) + 4	Use grouping symbols. You have a 3-shelf bookcase with 5 books on each shelf, plus 4 on the floor.
10	(24 × 6) – 5	Use grouping symbols. There are 6 classes. Each class needs 24 cupcakes. Five were eaten, this is subtraction.

Question	Answer	Detailed Explanation
11	$(7 \times 12) \div 4$	Product means multiply.
12	$(54 \div 2) - 17$	Remember "less than" means subtract. However, it says to subtract from the quotient, the answer you get from dividing.
13	$(210 + 14) \times 3$	Add 210 and 14; then multiply by 3.
14	$(8 \times 9) \div (1 + 3)$	You are dividing the product of two numbers by the sum of two numbers.
15	Actual answer is 6.	The parentheses indicate you should subtract 5 from 8 and then multiply by 2. Alejandro ignored the parentheses and multiplied 2 times 8 and then subtracted 5 to get an answer of 11.
16	$(12 \times 4) - 9$ = Nine less than the product of 12 and 4 $9 \times (12 - 4)$ = Four less than 12 multiplied by 9 $(12 - 9) \times 4$ = The difference between 12 and 9 multiplied by 4	

Generating Patterns and Identifying Relationships

THE STANDARD

CCSS 5.OA.3. Generate two numerical patterns using two given rules. Identify apparent relationships between corresponding terms. Form ordered pairs consisting of corresponding terms from the two patterns, and graph the ordered pairs on a coordinate plane. For example, given the rule "Add 3" and the starting number 0, and given the rule "Add 6" and the starting number 0, generate terms in the resulting sequences, and observe that the terms in one sequence are twice the corresponding terms in the other sequence. Explain informally why this is so.

What Does This Mean?

In math, there are times in which you are able to identify the relationship between the numbers based upon following a numerical pattern.

Example

Given the numbers 4, 7, 10, 13, . . . find the next three numbers.

It should be easy to identify that the next three numbers in the sequence are 16, 19, and 22. Why? Well, each number in the example above is 3 more than the number before it. So if you continue the pattern of "adding 3" you can figure out the next three, five, seven, and ten numbers in the pattern. You call the number pattern the rule. The rule for the number pattern above is "add 3."

You will learn to identify the rules of two numerical patterns and the relationships between the numbers in the sequences. Once you can identify the rule and the sequence, you will see that the two numbers written in a certain order form an **ordered pair** that can be graphed on the coordinate plane.

What Do We Mean When We Say Two Sets of Numbers Are Related Based upon the Rule?

Let's imagine that your art teacher needs to determine how many boxes of chalk to buy for a class of 12 students. She created the table below:

Students	0	3	6	9	12
Boxes	0	6	12	18	?

If you look at the first number sequence, *students*, you can see that the number sequence or pattern is increasing by three. So the rule for students is **"add 3."**

RULE: Add 3

Students	0	3	6	9	12
		+3	+3	+3	+3

If you look at the second number sequence, boxes, you can see that the number sequence or pattern is increasing by six. So the rule is **"add 6."**

RULE: Add 6

Boxes	0	6	12	18	?
		+6	+6	+6	+6

If your teacher needs to determine how many boxes to buy, she would add 6 to 18, which would give her 24. So for 12 students she needs to buy 24 boxes of chalk.

Look a little closer!

× 2

Rule + 3	Students	0	3	6	9	12
Rule + 6	Boxes	0	6	12	18	24

Let's look at the relationship between the students and the boxes of chalk. The boxes of chalk actually represent 2 times the number of students. In reality what you have is:

Students	0	3	6	9	12
Boxes	**2 × 0 = 0**	**2 × 3 = 6**	**2 × 6 = 12**	**2 × 9 = 18**	**2 × 12 = 24**

You can also refer to this table as an input-output table. The input in this case is students. When the mathematical rule, "multiply by 2," is applied to the input you get the output numbers as the result. The rule for the input-output table is **"multiply by 2."**

INPUT	Students	0	3	6	9	12
OUTPUT	**Boxes**	**2 × 0 = 0**	**2 × 3 = 6**	**2 × 6 = 12**	**2 × 9 = 18**	**2 × 12 = 24**

You can now represent this relationship as an ordered pair: (students, **boxes**).

The first number in your ordered pair represents the number of students and the second number represents the boxes of chalk.

The ordered pairs, also known as points, for this table would look like this:

(0, **0**), (3, **6**), (6, **12**), (9, **18**), (12, **24**)

These points can also be referred to as *x*, *y*-coordinate pairs.

Now that you have your ordered pairs, you can plot them on a graph (the coordinate plane). Note: You will learn more about the coordinate plane and graphing points in Chapter 7. However, for now let's take an introductory look. You are looking at the (*x*, *y*)-axes of the coordinate plane. Your input or the *x*-axis goes across the page, known as the horizontal axis. Your output or your *y*-axis runs up and down the page, known as the vertical axis.

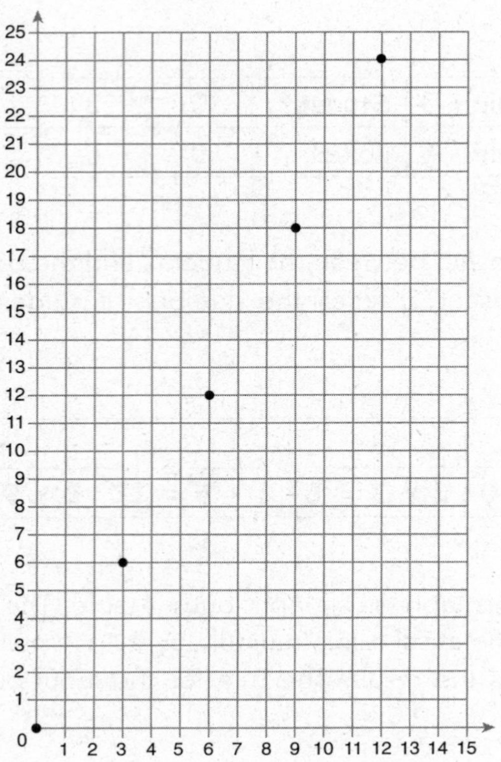

Look at the table below. You try!

INPUT	Hours worked	0	1	2	3	4
OUTPUT	Dollars earned	**0**	**4**	**8**	**12**	**?**

1. What is the rule for hours worked? ______________________________

Rule:	Hours worked	0	1	2	3	4

2. What is the rule for dollars earned? ______________________________

Rule:	Dollars earned	**0**	**4**	**8**	**12**	**?**

3. A. What is the relationship between the hours worked and dollars earned?

____

____

B. What is the rule for the input-output table? ____________________

4. How much money do you earn after working 4 hours? ____________________

5. Write your ordered pairs.

(___, ___), (___, ___), (___, ___), (___, ___), (___, ___)

6. Graph your ordered pairs on the coordinate plane.

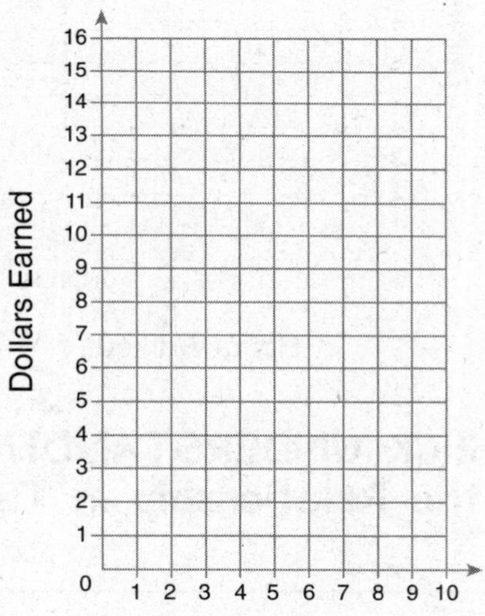

Answers (pages 120–121)

1. Hours worked rule: Add 1
2. Dollars earned rule: Add 4
3. **A.** The dollars earned represents 4 times the hours worked.
   B. The rule for the input-output table is multiply by 4.
4. $16
5. (0, 0), (1, 4), (2, 8), (3, 12), (4, 16)
6.

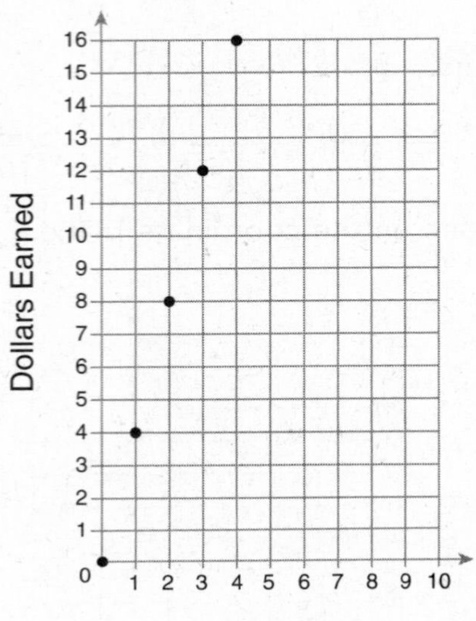

PRACTICE—CHECK UNDERSTANDING: Find the Rule to Determine the Relationship in Two Numerical Patterns

1. An input-output table has the rule "add 8." Given an input value of 3, what is the output value?

2. An input-output table has the rule "divide by 3." Given an input value of 15, what is the output value?

3. The local YMCA is signing up children for its volleyball clinic. The input-output table below shows the number of children that signed up and the number of volleyball teams that can be formed. How many teams can be formed if 54 students sign-up?

INPUT	Children	12	18	24	30
OUTPUT	Teams	**2**	**3**	**4**	**5**

○ A. 7 ○ B. 9 ○ C. 10 ○ D. 8

4. Use the input-output table in question 3. Which two coordinate pairs are values for the table?

- ☐ A. (6, 36)
- ☐ B. (42, 7)
- ☐ C. (66,11)
- ☐ D. (10, 60)

5. A input-output table has the rule, "subtract 7." Given an input value of 9, what is the output value?

- ○ A. 2
- ○ B. 63
- ○ C. 16
- ○ D. 0

6. Margaret wants to make an input-output table with the number of days in a week as the input and the number of hours in the days as the output. What is the rule for her input-output table?

○ A. Add 24 ○ B. Divide by 24 ○ C. Multiply by 7 ○ D. Multiply by 24

7. Lailah's father owns a shop that sells bicycles and wagons.

Part A: Create an input-output table using 1, 2, 3, 4, 5, 6, 7, and 8 as your input values. Determine the rule for the table to find your output values for the number of wheels needed for the bikes.

INPUT	Bike(s)								
OUTPUT	Wheel(s)								

RULE: ______________________________

Part B: Her brother who works in the shop makes the wagons. Use the same input values from Part A to create an input-output table for the wagons.

INPUT	Wagon(s)								
OUTPUT	Wheel(s)								

RULE: ______________________________

Part C: Graph the results from both tables on the graph below.

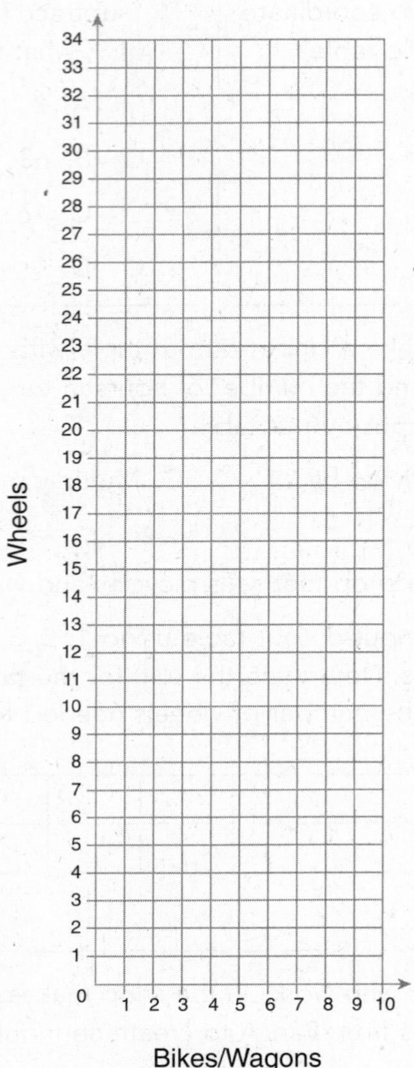

What is the relationship between the wheels needed for bicycles and wagons?

____

____

____

Answers (pages 122–124)

<table>
<tr><th>Question</th><th>Answer</th><th>Detailed Explanation</th></tr>
<tr><td>1</td><td>11</td><td>3 + 8 = 11</td></tr>
<tr><td>2</td><td>5</td><td>15 divided by 3 = 5</td></tr>
<tr><td>3</td><td>B</td><td>The rule is "divided by 6," 54 divided by 6 = 9.</td></tr>
<tr><td>4</td><td>B, C</td><td>Looking at the input table, the input value for children is getting larger not smaller. The x values for B and C divided by 6 will give you the number of teams that can be formed.</td></tr>
<tr><td>5</td><td>A</td><td>9 – 7 = 2</td></tr>
<tr><td>6</td><td>D</td><td>There are 24 hours in a day. 2 days = 48 hours, 2 × 24 = 48.</td></tr>
<tr><td>7</td><td colspan="2">See Detailed Explanation.
Part A:
<table>
<tr><td>Bikes</td><td>1</td><td>2</td><td>3</td><td>4</td><td>5</td><td>6</td><td>7</td><td>8</td></tr>
<tr><td>Wheels</td><td>2</td><td>4</td><td>6</td><td>8</td><td>10</td><td>12</td><td>14</td><td>16</td></tr>
</table>
Rule: Multiply by 2.
Part B:
<table>
<tr><td>Wagons</td><td>1</td><td>2</td><td>3</td><td>4</td><td>5</td><td>6</td><td>7</td><td>8</td></tr>
<tr><td>Wheels</td><td>4</td><td>8</td><td>12</td><td>16</td><td>20</td><td>24</td><td>28</td><td>32</td></tr>
</table>
Rule: Multiply by 4.</td></tr>
</table>

Question	Answer	Detailed Explanation
	Part C:	
		Wheels Bikes/Wagons Bikes ○ Wagons ●
		The wagons require two times more wheels than the bicycles.

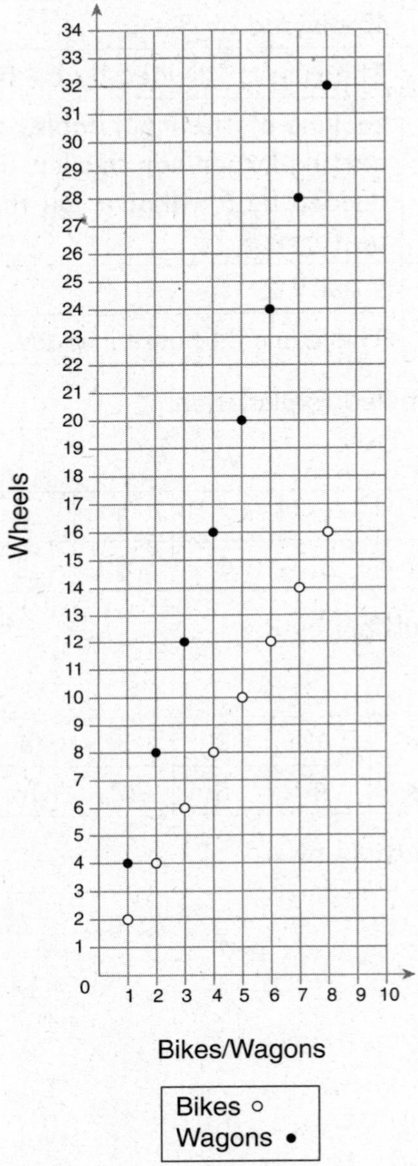

Number and Operations—Fractions

CHAPTER 5

You might be thinking, "Why the push to learn fractions?" Have you ever had to divide something between you and others? Help your parents leave the proper tip in a restaurant? Want to know the batting average or shooting percentage of your favorite athlete? And just what does 15% mean anyhow?

If any—or all—of those questions have ever crossed your mind, then you have wanted to use fractions and perhaps didn't even know it. Fractions are something we use often without thinking. Mastering fractions is a skill that will help you in everyday activities and, even better, will provide a base that will help you master algebra and more advanced mathematics in the grades to come.

Thinking about fractions and even working with fractions can be challenging. However, having a solid understanding of fractions will take you a long way. Your ability to think and reason with fractions means that you can recognize relationships and make comparisons between quantities.

The writers of the Common Core Standards recognize that developing an understanding of fractions and fractional relationships happens over time and across grade levels. Thus, in fifth grade, you will use your knowledge of fractions and fraction models learned in fourth grade to represent the addition and subtraction of fractions with unlike denominators as equivalent calculations with like denominators. You will develop fluency in calculating sums and differences of fractions in addition to being able to make realistic estimates with fractions. Additionally, you will use your knowledge of multiplication and division (thus far of whole numbers) to understand and make sense of what multiplying and dividing fractions really means.

So push to learn fractions!

VISUALIZE THE MATH

1. Look at the models below. What fraction of each rectangle is shaded gray? Write your answer on the lines provided below.

A.

B.

______ ______

2. Using the shapes below draw a picture to create equivalent fractions for each shape. Note: The fraction for each shape must have the same denominator.

HELPFUL HINT

PARCC TEST STRATEGY: Draw visual fraction models to show equivalence.

Show your work here:	Show your work here:

KEY CONCEPT

Equivalent fractions have the same value because they represent the same number and have equal amounts.

Answers

1. **A.** $\frac{1}{2}$ **B.** $\frac{2}{3}$

2. Denominator for each picture should be six. Pictures should be $\frac{3}{6}$ and $\frac{4}{6}$.

Using Equivalent Fractions as a Strategy to Add and Subtract Fractions

THE STANDARDS

CCSS.5.NF.1. Add and subtract fractions with unlike denominators (including mixed numbers) by replacing given fractions with equivalent fractions in such a way as to produce an equivalent sum or difference of fractions with like denominators. For example, $\frac{2}{3}+\frac{5}{4}=\frac{8}{12}+\frac{15}{12}=\frac{23}{12}$.

$$\left(\text{In general, } \frac{a}{b}+\frac{c}{d}=\frac{(ad+bc)}{bd}.\right)$$

CCSS.5.NF.2. Solve word problems involving addition and subtraction of fractions referring to the same whole, including cases of unlike denominators (e.g., by using visual fraction models or equations to represent the problem). Use benchmark fractions and number sense of fractions to estimate mentally and assess the reasonableness of answers. For example, recognize an incorrect result $\frac{2}{5}+\frac{1}{2}=\frac{3}{7}$, by observing that $\frac{3}{7}<\frac{1}{2}$.

What Does This Mean?

To add and subtract fractions with unlike denominators (including mixed numbers), you need to have common denominators.

HELPFUL HINT

In fourth grade, you learned to create equivalent fractions and to compare two fractions with different denominators by creating common denominators. Remember that multiplying the denominators will create a common denominator. *Note*: If you use this strategy to add and subtract fractions, the common denominator may not be in the *simplest form*.

Examples

$$\frac{1}{5}+\frac{3}{8}=\frac{1\times 8}{5\times 8}+\frac{3\times 5}{8\times 5}=\frac{8}{40}+\frac{15}{40}=\frac{23}{40}$$

$$2\frac{1}{4}-\frac{1}{6}=2\frac{6}{24}-\frac{4}{24}=2\frac{2}{24}=2\frac{1}{12}$$

PRACTICE—CHECK UNDERSTANDING: Common Denominators and Equivalent Fractions

Write an equivalent fraction for each fraction pair below by finding the common denominator:

Fraction Pair	Common Denominator	Equivalent Fractions
1. $\frac{2}{8}, \frac{1}{6}$	8 × 6 = 48	$\frac{12}{48}, \frac{8}{48}$
2. $\frac{3}{4}, \frac{3}{8}$		,
3. $\frac{2}{5}, \frac{3}{10}$		,
4. $\frac{1}{3}, \frac{5}{8}$		,

Answers (page 130)

Question	Answer
1	Done for you
2	$32, \frac{24}{32}, \frac{12}{32}$
3	$50, \frac{20}{50}, \frac{15}{50}$
4	$24, \frac{8}{24}, \frac{15}{24}$

Common denominator—A common denominator of two fractions is a number that is a common multiple of the denominators.

Strategy 1

Add $\frac{1}{4} + \frac{2}{5}$. We need to find the common denominator of $\frac{1}{4}$ and $\frac{2}{5}$.

Step 1: Identify the denominators.

Denominators are 4 and 5.

Step 2: Multiply the numerator and the denominator of each fraction by the denominator of the other fraction.

- $\frac{1}{4} = \frac{1 \times 5}{4 \times 5} = \frac{5}{20}$ Equivalent fractions
- $\frac{2}{5} = \frac{2 \times 4}{5 \times 4} = \frac{8}{20}$

Common denominator is 20.

Step 3: Rewrite the problem using the equivalent fractions and add.

$$\frac{5}{20} + \frac{8}{20} = \frac{13}{20}$$

Strategy 2

Subtract $\frac{1}{4} - \frac{1}{6}$. Find the common denominator of $\frac{1}{4}$ and $\frac{1}{6}$.

Step 1: Write the equivalent fractions for $\frac{1}{4}$ and $\frac{1}{6}$.

$$\frac{1}{4}: \frac{1}{4}, \frac{2}{8}, \frac{3}{12}$$

$$\frac{1}{6}: \frac{1}{6}, \frac{2}{12}, \frac{3}{18}$$

Step 2: Circle the two fractions with the same denominator.

$$\mathbf{\frac{1}{4}}: \frac{1}{4}, \frac{2}{8}, \boxed{\frac{3}{12}}$$

$$\mathbf{\frac{1}{6}}: \frac{1}{6}, \boxed{\frac{2}{12}}, \frac{3}{18}$$

Step 3: Rewrite the problem using the equivalent fractions and subtract.

$$\frac{3}{12} - \frac{2}{12} = \frac{1}{12}$$

KEY CONCEPT

When adding and subtracting fractions, the fractions must have the same denominator.

PRACTICE—CHECK FOR UNDERSTANDING: Adding and Subtracting Fractions

1. $\frac{2}{3} + \frac{1}{6} =$ ______

2. $\frac{1}{5} + \frac{3}{10} =$ ______

3. $\frac{5}{8} - \frac{1}{4} =$ ______

4. $\frac{5}{9} + \frac{2}{8} =$ ______

5. $\frac{7}{8} - \frac{1}{2} =$ ______

6. $\frac{5}{12} - \frac{1}{3} =$ ______

7. $\frac{3}{4} + \frac{2}{5} - \frac{1}{10} =$ ______

PRACTICE—CHECK FOR REASONING: Adding and Subtracting Fractions (Proper Fractions)

8. Ryan added the fractions below and got an answer of $\frac{3}{7}$.

$$\frac{1}{4}+\frac{2}{3}=\_\_\_\_\_$$

- Use your knowledge of adding fractions to explain why Ryan's answer is incorrect.

____

____

____

____

- What is the correct answer? Explain how you arrived at your answer in the box below.

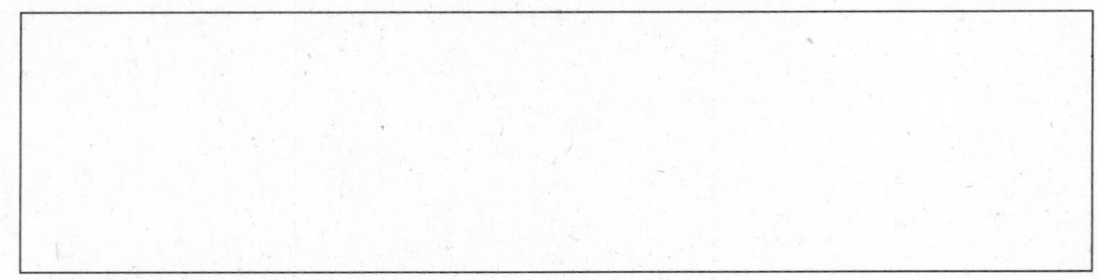

9. Jordan says that $\frac{9}{10}-\frac{4}{6}=\frac{9}{10}-\frac{1}{3}-\frac{1}{3}$. Use your knowledge of adding and subtracting fractions to explain Jordan's reasoning. You can include numbers and/or words in your explanation.

____

____

____

____

10. Use your knowledge of adding and subtracting fractions to explain how two unlike fractions can have a sum of 1. You can include numbers and/or words in your explanation.

____

____

____

____

PRACTICE—CHECK FOR APPLICATION: Adding and Subtracting Fractions (Proper Fractions)

11. Jorge wants to sell lemonade at the school fair. The principal, Ms. Warren, informed Jorge that all vendors are required to contribute $\frac{1}{4}$ of all sales to the school and also pay the PTA $\frac{1}{5}$ of all sales for a booth rental.

Part A: What fraction of the sales does Jorge pay the school and the PTA?

Part B: What fraction of the sales does Jorge get to keep after paying the required expenses?

12. Jessica is taking an Arts & Crafts class and has to complete a final project that requires her to use 1 yard of fabric. The table below shows the different items Jessica can make.

Item	Yards of Fabric Needed
Seat Cover	$\frac{1}{3}$
Puppet	$\frac{1}{6}$
Dolls Dress	$\frac{1}{8}$
Pillow	$\frac{1}{2}$

- Which items can Jessica make that will use exactly 1 yard of fabric? Show your work in the space below:

- Explain how you arrived at your answer.

Answers (pages 133–136)

Question	Answer	Detailed Explanation
1	$\frac{5}{6}$	Remember when adding fractions, the fractions must have the same denominator. Using the algorithm $\frac{a}{b}+\frac{c}{d}=\frac{ad}{bd}+\frac{cb}{bd}$, you would have the following answer. $\frac{2}{3}+\frac{1}{6}=\frac{2\times 6}{3\times 6}+\frac{1\times 3}{6\times 3}=\frac{12}{18}+\frac{3}{18}=\frac{15}{18}=\frac{5}{6}$
		Using equivalent fractions, you would have the following answer. **Step 1:** $\mathbf{\frac{2}{3}}: \frac{4}{6}, \frac{6}{9}, \frac{8}{12}, \frac{10}{15}, \frac{12}{18}$ (circled) $\mathbf{\frac{1}{6}}: \frac{1}{6}, \frac{2}{12}, \frac{3}{18}$ (circled) **Step 2:** Add $\frac{12}{18}+\frac{3}{18}=\frac{15}{18}$ **Step 3:** Simplify $\frac{15}{18}=\frac{5}{6}$
2	Common denominator = 10 Answer: $\frac{5}{10}=\frac{1}{2}$	
3	Common denominator = 32 Answer: $\frac{12}{32}=\frac{3}{8}$	
4	Common denominator = 72 Answer: $\frac{58}{72}=\frac{29}{36}$	
5	Common denominator = 16 Answer: $\frac{6}{16}=\frac{3}{8}$	

Question	Answer	Detailed Explanation
6	Common denominator = 36 Answer: $\frac{3}{36}=\frac{1}{12}$	
7	$1\frac{1}{20}$ or $\frac{21}{20}$	Using equivalent fractions to find the common denominator = 20 **Remember we have to solve the problem from left to right, no parentheses here.** $\frac{15}{20}+\frac{8}{20}-\frac{2}{20}=\frac{23}{20}-\frac{2}{20}=\frac{21}{20}$ *Note*: This is an improper fraction. ↗
8	The correct answer should be $\frac{11}{12}$. I used the strategy of multiplying the denominators to find the common denominator.	Fractions need to have denominators with the same value, called the common denominator. Ryan just added the numerators and the denominators instead of multiplying the denominators or using equivalent fractions to find the common denominator.
9		Well, Jordan knows that $\frac{1}{3}=\frac{2}{6}$. So $\frac{1}{3}+\frac{1}{3}=\frac{2}{6}+\frac{2}{6}$. Jordan recognizes that $\frac{4}{6}=\frac{2}{3}$. Jordan realizes the common denominators are 30 and 60. He understands that 30 and 60 are multiples of 3, 6, and 10.
10		A fraction is a way of representing part of a whole. Not all fractions have the number of parts. This means they have different denominators. The picture shows $\frac{1}{2}=\frac{2}{4}$ so if I add $\frac{1}{2}+\frac{2}{4}=\frac{2}{4}+\frac{2}{4}=\frac{4}{4}=1$ whole. =

Question	Answer	Detailed Explanation
11	**Part A:** $\frac{1}{4}+\frac{1}{5}=\frac{5}{20}+\frac{4}{20}=\frac{9}{20}$ **Part B:** Jorge gets $\frac{11}{20}$ of the sales after paying cost.	The whole is $\frac{20}{20}=1$. To find how much Jorge keeps, I have to subtract. $1-\frac{9}{20}=\frac{20}{20}-\frac{9}{20}=\frac{11}{20}$
12	Jessica can make the seat cover, the puppet, and the pillow.	The fractions must have a sum of 1 yard = 1 whole. $\frac{1}{3}+\frac{1}{6}+\frac{1}{2}=\frac{2}{6}+\frac{1}{6}+\frac{3}{6}=\frac{6}{6}=1$. Any other combination of items is a fraction that is less than 1.

VISUALIZE THE MATH

What is a mixed number or improper fraction?

The picture below shows 3 squares. Each square is equally divided into 4 parts.

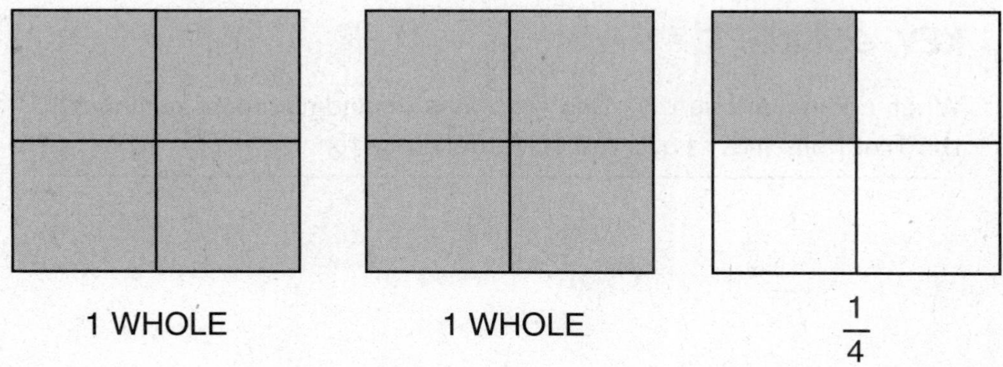

What fraction of the 3 squares is shaded?

$$2\frac{1}{4} = \frac{9}{4}$$

↑ Mixed number ↑ Improper fraction

A **mixed number fraction** is nothing more than the whole number and the proper fraction.

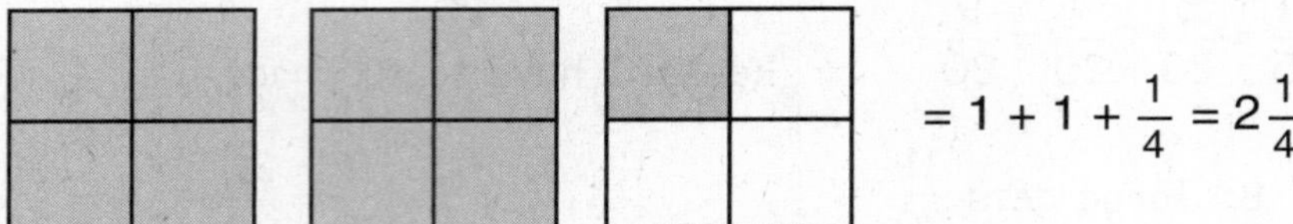

$= 1 + 1 + \frac{1}{4} = 2\frac{1}{4}$

An **improper fraction** is nothing more than a fraction that has a numerator that is greater than the denominator.

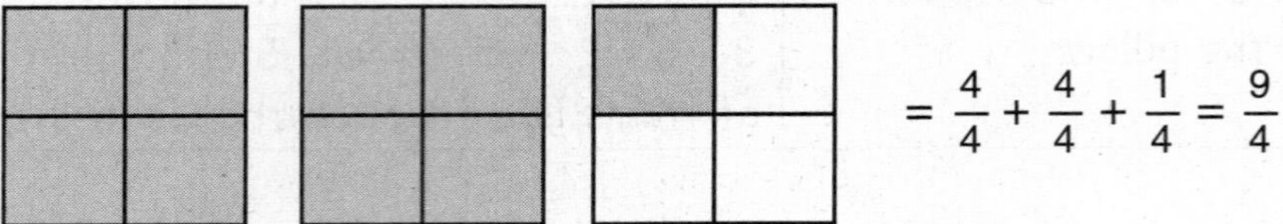

$= \frac{4}{4} + \frac{4}{4} + \frac{1}{4} = \frac{9}{4}$

Add $3\frac{2}{3} + \frac{3}{5}$

$$= 3\frac{10}{15} + \frac{9}{15} = 3\frac{19}{15} = 3 + \frac{15}{15} + \frac{4}{15} = 4\frac{4}{15}$$

KEY CONCEPT

When adding and subtracting fractions (including mixed numbers), the fractions must have the same denominator.

PRACTICE—CHECK FOR UNDERSTANDING: Adding and Subtracting (Mixed Number or Improper Fractions)

1. $4\frac{2}{5}+\frac{3}{4}=$ ______

2. $\frac{12}{3}+\frac{1}{5}=$ ______

3. $5\frac{1}{8}-\frac{1}{2}$ ______

4. $\frac{11}{9}+\frac{4}{3}=$ ______

5. $3\frac{7}{8}-\frac{4}{2}=$ ______

6. $7\frac{5}{12}-5\frac{1}{3}=$ ______

7. $4\frac{1}{6}+\frac{7}{3}-\frac{3}{4}=$ ______

PRACTICE—CHECK FOR REASONING: Adding and Subtracting Fractions (Mixed Number and Improper Fractions)

8. Jessica is making pumpkin gelato for her and 5 friends using the recipe below.

Pumpkin Gelato Ingredients*

$2\frac{2}{3}$ cups pumpkin puree

$2\frac{3}{4}$ cups heavy cream

$1\frac{3}{4}$ cups whole milk

$1\frac{1}{2}$ cups honey

1 teaspoon cinnamon

$\frac{1}{2}$ teaspoon nutmeg

Pinch of salt

***A recipe does exist for pumpkin gelato. However, the recipe above is not accurate and was created just for this math problem.**

Directions

Whisk together the pumpkin puree, cream, milk, sugar, cinnamon, nutmeg, and salt in a large bowl. Freeze the mixture in an ice cream maker according to the manufacturer's directions.

Part A: Jessica's bowls hold a $\frac{1}{2}$ cup of gelato. Explain how you can determine if the recipe makes enough for her friends and her to have two bowls of gelato each. You can include numbers and/or words in your explanation.

Part B: Three friends decided to have only one bowl of gelato. How much gelato is left if everyone else has two bowls?

Show your work in the space below:

Answers (pages 141–143)

Question	Answer	Detailed Explanation
1	$5\frac{3}{20}$	$4\frac{8}{20}+\frac{15}{20}=4\frac{23}{20}=4+\frac{20}{20}+\frac{3}{20}=5\frac{3}{20}$
2	$4\frac{3}{15}=4\frac{1}{5}$	
3	$4\frac{5}{8}$	$5\frac{2}{16}-\frac{8}{16}=4\frac{18}{16}-\frac{8}{16}=4\frac{10}{16}=4\frac{5}{8}$
4	$2\frac{5}{9}$	$\frac{33}{27}+\frac{36}{27}=\frac{69}{27}=\frac{27}{27}+\frac{27}{27}+\frac{15}{27}=2\frac{15}{27}=2\frac{5}{9}$
5	$1\frac{7}{8}$	$3\frac{7}{8}-\frac{4}{2}=3\frac{14}{16}-\frac{32}{16}=1\frac{46}{16}-\frac{32}{16}=1\frac{14}{16}=1\frac{7}{8}$
6	$2\frac{1}{12}$	$7\frac{5}{12}-5\frac{1}{3}=7\frac{15}{36}-5\frac{12}{36}=2\frac{3}{36}=2\frac{1}{12}$
7	$5\frac{3}{4}$	Remember the order of operations. $4\frac{1}{6}+\frac{7}{3}=4\frac{3}{18}+\frac{42}{18}=4\frac{45}{18}=$ $4+\frac{18}{18}+\frac{18}{18}+\frac{9}{18}=6\frac{9}{18}=6\frac{1}{2}$ $6\frac{1}{2}-\frac{3}{4}=6\frac{4}{8}-\frac{6}{8}=5\frac{12}{8}-\frac{6}{8}=5\frac{6}{8}=5\frac{3}{4}$
8	**Part A:** Yes	Possible answer: If you just add the puree, cream, milk, and honey you have $8\frac{2}{3}$ cups. I did not add the spices because a teaspoon is a very small amount compared to a cup. Since two bowls equals 1 cup. There are 6 people, so you would need 6 cups. You have enough for everyone to have two bowls.
	Part B: See Detailed Explanation.	Using the same information you have $8\frac{2}{3}$ cups. Three people have two bowls that represents 3 cups. Three people have one bowl that represents $\frac{1}{2}+\frac{1}{2}+\frac{1}{2}=1\frac{1}{2}$. All 6 people eat $4\frac{1}{2}$ cups $\left(3+1\frac{1}{2}\right)$ of gelato, leaving $4\frac{1}{6}$ cups remaining $\left(8\frac{2}{3}-4\frac{1}{2}\right)$.

Multiplying Fractions (by Whole Numbers and Fractions)

THE STANDARDS

CCSS.5.NF.4. Apply and extend previous understandings of multiplication to multiply a fraction or whole number by a fraction.

a. Interpret the product $\left(\frac{a}{b}\right)\times q$, as *a* parts of a partition of *q* into *b* equal parts; equivalently, as the result of a sequence of operations $a \times q \div b$. For example, use a visual fraction model to show $\left(\frac{2}{3}\times 4 = \frac{8}{3}\right)$, and create a story context for this equation. Do the same with $\left(\frac{2}{3}\right)\times\left(\frac{4}{5}\right)=\frac{8}{15}$. In general, $\frac{a}{b}\times\frac{c}{d}=\frac{ac}{bd}$.

b. Find the area of a rectangle with fractional side lengths by tiling it with unit squares of the appropriate unit fraction side lengths, and show that the area is the same as would be found by multiplying the side lengths. Multiply fractional side lengths to find areas of rectangles, and represent fraction products as rectangular areas.

CCSS.5.NF.6. Solve real-world problems involving multiplication of fractions and mixed numbers (e.g., by using visual fraction models or equations to represent the problem).

What Does This Mean?

You know multiplication can be represented as repeated addition. Additionally, you learned how to model multiplication using pictures and area models. For example, if Samantha made 6 pillows per day in sewing class for 3 days; you realize it can be written as 6 + 6 + 6 = 3 × 6 = 18 or you can model the situation as the picture

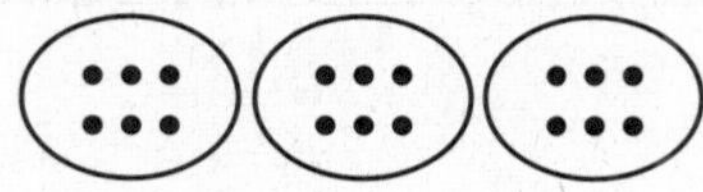

Even though you will now be multiplying a whole number by a fraction, the concept of multiplication remains the same. Imagine that you have committed to run $\frac{1}{4}$ mile every day for the next 5 days. This could be written as $\frac{1}{4}+\frac{1}{4}+\frac{1}{4}+\frac{1}{4}+\frac{1}{4}=\frac{5}{4}$ miles or $\frac{1}{4}\times 5=(1\times 5)\div 4=1\frac{1}{4}$. Additionally, you can model it as

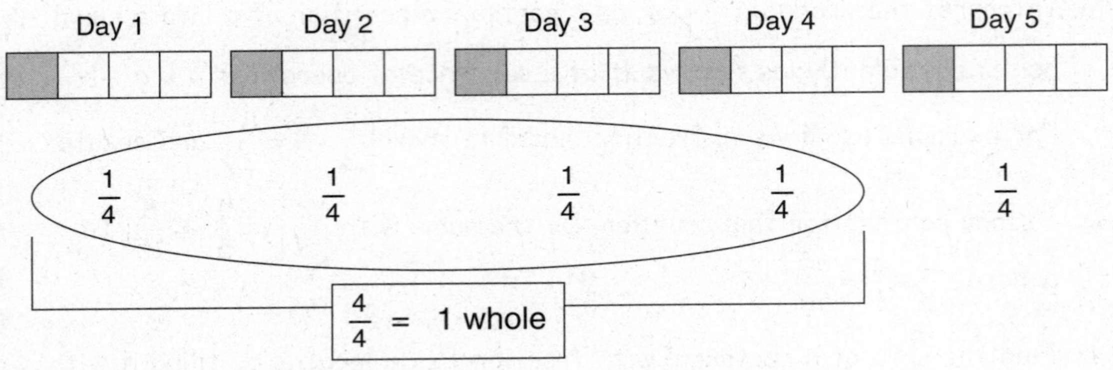

The general rule of thumb or formula for multiplying a whole number by a fraction is:

$$\left(\frac{a}{b}\times q=\frac{a\times q}{b}\right)$$

$$b\neq 0$$

↑

b cannot equal zero

KEY CONCEPT

Remember the commutative property of multiplication

$$\frac{a}{b}\times q=q\times\frac{a}{b}$$

Another Strategy

If you had 35 books in your family library and you read $\frac{2}{5}$ of them over the summer, exactly how many books did you read?

Step 1: Set up your equation.

$$\frac{\boxed{2}}{⑤} \times 35 = ?$$

HELPFUL HINT

You are separating the 35 books into ⑤ equal parts (35 ÷ 5 = ?); however, you only read 2 groups of the whole (35 books).

Step 2: Draw a rectangular area model to show the 35 books separated into 5 equal parts.

(35 ÷ 5 = 7), each equal part has 7 books

5 Equal Parts

$\frac{1}{5}$	1	2	3	4	5	6	7
$\frac{2}{5}$							
$\frac{3}{5}$							
$\frac{4}{5}$							
$\frac{5}{5}$							

Step 3: Shade in the squares for 2 groups of the 5 parts. Count the number of shaded squares. You have 14 shaded squares.

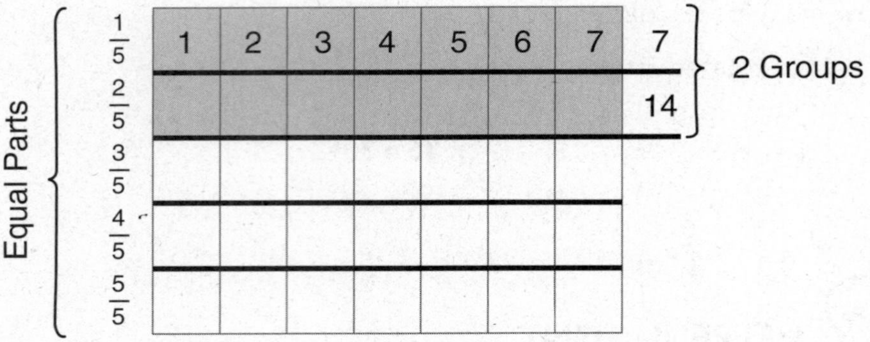

Equal Parts								
$\frac{1}{5}$	1	2	3	4	5	6	7	7
$\frac{2}{5}$								14
$\frac{3}{5}$								
$\frac{4}{5}$								
$\frac{5}{5}$								

2 Groups

Using the formula, you would multiply the numerator by the whole number and divide the product of the two numbers by the denominator.

$$\frac{2}{5}\times 35=\frac{(2\times 35)}{5}=\frac{70}{5}=14 \leftarrow \text{You read 14 books}$$

Let's look at using our math knowledge to find another way to solve this problem.

What if you looked at decomposing the numbers, to find common factors, before trying to solve the problem? Then you could rewrite the problem as:

$$\frac{2}{5}\times 35=\frac{2}{5}\times 7\times 5=\frac{2\times 7\times 5}{1\times 5}$$

Well, since you know $\frac{5}{5}=1$ whole, you can rewrite the equation as:

$$\frac{2\times 7}{1}=\frac{14}{1}=14$$

I think you are ready to give some problems a try.

PRACTICE—CHECK FOR UNDERSTANDING: Multiplying Whole Numbers by Fractions

1. $\frac{2}{3}\times 12=$ ______

2. $\frac{3}{8}\times 6=$ ______

3. $20\times\frac{4}{5}=$ ______

4. $13\times\frac{3}{4}=$ ______

5. $\frac{7}{8}\times 16=$ ______

6. $\frac{5}{12}\times 8=$ ______

7. A professional basketball team played 108 games last season. They won $\frac{1}{4}$ of their games in overtime and lost only $\frac{1}{6}$ of their games all season. Record the team's win-loss record below.

Games: Won in overtime ______________	Lost: ______________
Show your work:	Show your work:

8. Mr. Maxwell owns a small landscaping company. He has snow blowers and a snow plow. The snow blowers' gas tank holds a total of 15 gallons of gas and the snow plow's gas tank holds 24 gallons of gas. He used $\frac{4}{5}$ of a tank of gas for all the snow blowers and $\frac{5}{8}$ of a tank of gas for the snow plow. Determine how many total gallons of gas Mr. Maxwell used.

Provide your answer in the space below.

9. Which statement can be represented by the expression $\frac{3}{8} \times 9$?

○ A. $\frac{3}{8} \times 9$ is 8 groups of 9, divided into 3 equal parts.

○ B. $\frac{3}{8} \times 9$ is 3 groups of 9, divided into 72 equal parts.

○ C. $\frac{3}{8} \times 9$ is 3 groups of 9, divided into 8 equal parts.

○ D. $\frac{3}{8} \times 9$ is 8 groups of 9, divided into 27 equal parts.

10. A recipe says that a $\frac{2}{3}$ cup of sugar is needed to make 1 gallon of lemonade.

Which of the following measurements represents the number of cups needed to make 9 gallons of lemonade?

○ A. $3\frac{1}{2}$

○ B. $2\frac{1}{4}$

○ C. 5

○ D. 6

11. Draw a rectangle to model the problem $\frac{3}{4} \times 8 = 6$.

Answers (pages 149–151)

Question	Answer	Detailed Explanation
1	8	$\frac{2\times12}{3}=\frac{24}{3}=8$
2	$\frac{18}{8}=\frac{9}{4}=2\frac{1}{4}$	
3	16	
4	$9\frac{3}{4}$	
5	14	
6	$\frac{10}{3}=3\frac{1}{3}$	
7	27 wins in overtime, 18 losses	$\frac{1\times108}{4}=27,\ \frac{1\times108}{6}=18$
8	27 gallons	Snow blowers $\frac{4}{5}\times15=12$
		Snow plow $\frac{5}{8}\times24=15$
9	C	$\frac{3\times9\text{ (three groups of 9)}}{8\text{ (divided by 8)}}$
10	D	$\frac{2\times9}{3}=\frac{18}{3}=6$
11		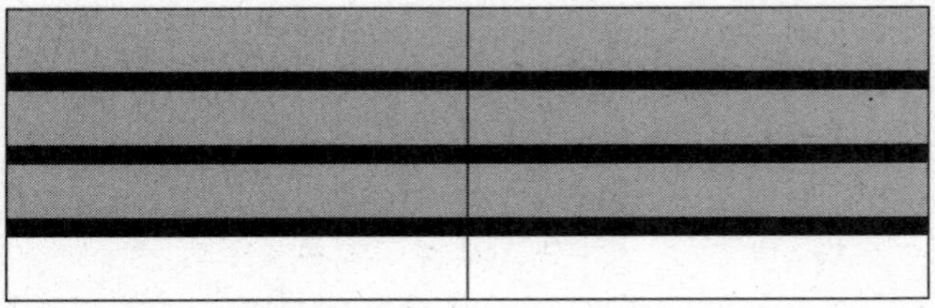 You are separating the 8 into 4 equal parts (8 ÷ 4 = 2). Each part has 2. You then need 3 groups of the whole which is 3 × 2 = 6.

You seem to understand how to multiply a whole number by a fraction. However, what happens when you multiply a fraction by a fraction? It may be a little harder to see multiplication of a fraction by another fraction without explaining the process using a rectangular area model first. Let's take for example that you are selling brownies at the school bake sale. You have $\frac{2}{3}$ of a pan of brownies left and a friend's mother wants to buy $\frac{3}{4}$ of the remaining brownies. What fraction of the reaming brownies does she actually buy?

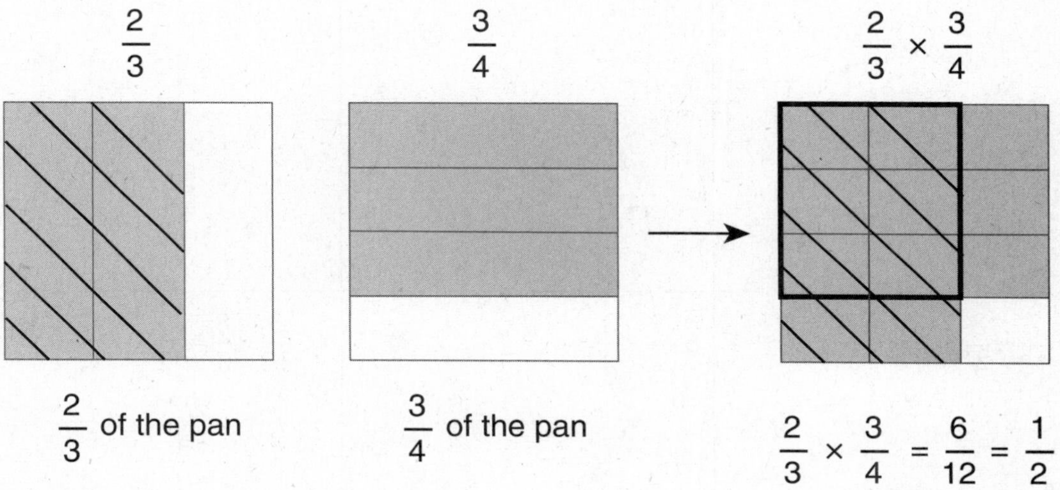

What happened here? You had 1 whole pan of brownies that you cut into 3 equal parts or *thirds*. You then took what you had left, $\frac{2}{3}$, and cut that into four equal parts or *fourths*. You realize that you have 12 equal parts or *twelfths*. Your friend's mother actually buys $\frac{6}{12}$ of the remaining brownies. You can also say, she bought $\frac{1}{2}$ of the original pan of brownies.

The general rule of thumb or formula for multiplying a fraction by a fraction is:

$$\frac{a}{b} \times \frac{c}{d} = \frac{ac}{bd}$$
$$b, d \neq 0$$
$$\uparrow$$

b and/or *d* cannot equal zero

Let's try some more problems.

PRACTICE—CHECK FOR UNDERSTANDING: Multiplying Fractions by Fractions

1. $\frac{2}{5} \times \frac{5}{8} =$ ______

2. $\frac{3}{8} \times \frac{4}{9} =$ ______

3. $\frac{3}{12} \times \frac{4}{5} =$ ______

4. $\frac{7}{10} \times \frac{3}{4} =$ ______

5. $\frac{7}{8} \times \frac{3}{2} =$ ______

6. $\frac{5}{12} \times \frac{8}{3} =$ ______

7. Your class planted roses in $\frac{5}{6}$ of the school garden. When the roses bloomed in the spring $\frac{1}{3}$ were pink. The school garden is modeled below.

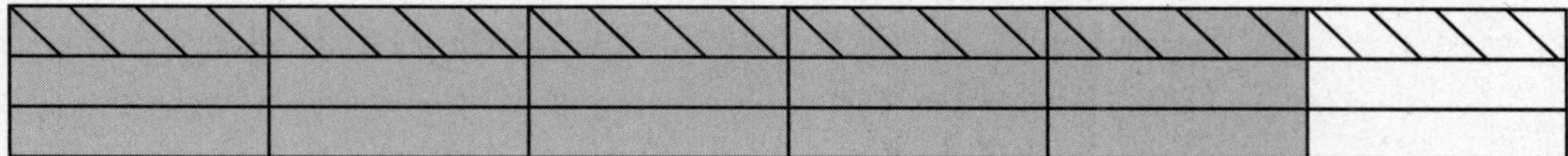

What part of the garden has pink roses?

○ A. $\frac{5}{10}$ ○ B. $\frac{1}{18}$

○ C. $\frac{5}{18}$ ○ D. $\frac{1}{3}$

8. Use the space below to draw a rectangle to model the problem $\frac{2}{3} \times \frac{1}{4} = \frac{2}{12}$.

9. Use the space below to draw a rectangle to model the problem $\frac{4}{5} \times \frac{1}{3} = \frac{4}{15}$.

10. Use the figure below.

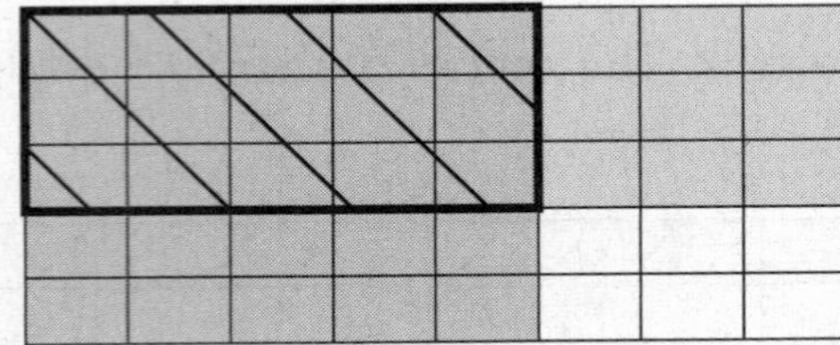

- What fraction multiplication problem is represented by the rectangular model above?
- What is the product of the fraction multiplication problem?

Show your work in the box below.

11. Find the product of the fraction multiplication problem represented below.

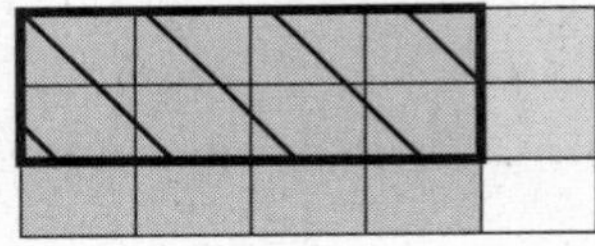

Show your work in the box below.

Answers (pages 154–156)

Question	Answer	Detailed Explanation
1	$\frac{10}{40}=\frac{1}{4}$	
2	$\frac{12}{72}=\frac{6}{36}=\frac{1}{6}$	$\frac{12}{72}=\frac{6\times 2}{36\times 2}=\frac{6}{36}=\frac{1\times 6}{6\times 6}=\frac{1}{6}$
3	$\frac{12}{60}=\frac{1}{5}$	
4	$\frac{21}{40}$	
5	$\frac{21}{16}=1\frac{5}{16}$	
6	$\frac{40}{36}=\frac{10}{9}=1\frac{1}{9}$	
7	*C*	$\frac{5}{6}\times\frac{1}{3}=\frac{5}{18}$
8	$\frac{2}{12}$ See Detailed Explanation.	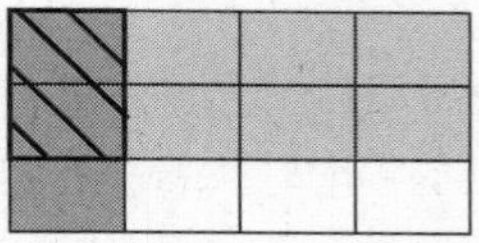
9	$\frac{4}{15}$ See Detailed Explanation.	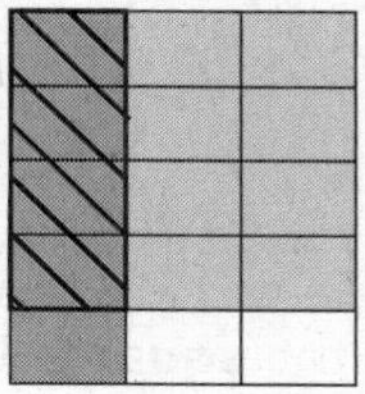
10	$\frac{3}{5}\times\frac{5}{8}=\frac{15}{40}=\frac{3}{8}$	$\frac{15}{40}=\frac{3\times 5}{8\times 5}=\frac{3}{8}$
11	$\frac{2}{3}\times\frac{4}{5}=\frac{8}{15}$	

Multiplying Fractions (Mixed Numbers)

Multiplying mixed numbers may seem like a difficult task. However, you have learned several strategies that can be applied to assist you in accomplishing this task. Let's start with a simple area problem using whole numbers and work our way into fractions.

You want to tile the wall behind the kitchen sink. The wall has the measurements of 2 feet × 1 foot. How many *square feet* of tile are needed?

KEY CONCEPT

The standard unit of measurement used to measure area is square units.

Using a rectangle to model the problem we get the following picture.

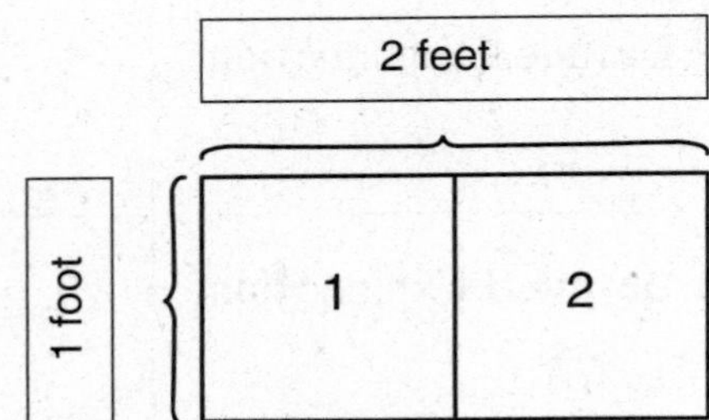

Counting the **squares** inside the figures you get 2. The area of the figure is 2 square feet. Now, what if the measurements of the wall behind the kitchen sink were $2\frac{2}{3}$ feet × $1\frac{1}{4}$ feet? You know that you are counting the squares within your figure; however, you are using mixed numbers this time. How do you approach solving this type of problem using a similar model? You are going to look at two strategies that will help you reach the same result.

Strategy 1

You need to find the area of $2\frac{2}{3}$ feet $\times 1\frac{1}{4}$ feet.

Step 1: $2\frac{2}{3}$ feet $\times 1\frac{1}{4}$ feet

You can rewrite each mixed number as the sum of the whole number and the fraction. Remember you did this in the beginning of the chapter.

$$2\frac{2}{3} = 2 + \frac{2}{3} \qquad\qquad 1\frac{1}{4} = 1 + \frac{1}{4}$$

Step 2: Draw a rectangular area model to show the rewritten mixed number fractions and find the area of each section.

	2	$+ \frac{2}{3}$
1	$2 \times 1 = 2$	$1 \times \frac{2}{3} = \frac{2}{3} = \frac{2}{3}$
$+ \frac{1}{4}$	$\frac{1}{4} \times 2 = \frac{2}{4}$	$\frac{1}{4} \times \frac{2}{3} = \frac{2}{12}$

Step 3: Add the areas of each section to find the total area of the rectangle.

$$2 + \frac{2}{3} + \frac{2}{4} + \frac{2}{12}$$
$$= 2 + \frac{8}{12} + \frac{6}{12} + \frac{2}{12}$$
$$= 2 + \frac{16}{12} = 2 + 1 + \frac{4}{12} = 3\frac{1}{3}$$

Step 4: State your answer.

$$2\frac{2}{3} \text{ feet} \times 1\frac{1}{4} \text{ feet} = 3\frac{1}{3} \text{ square feet}$$

You need $3\frac{1}{3}$ square feet of tile for the wall behind the sink.

Strategy 2

You need to find the area of $2\frac{2}{3}$ feet $\times 1\frac{1}{4}$ feet.

Step 1: Change both fractions to improper fractions.

$$2\frac{2}{3}=\underbrace{\frac{3}{3}+\frac{3}{3}}+\frac{2}{3}=\frac{8}{3} \quad \rightarrow \quad 2\frac{2}{3}=\frac{(2\times3)+2}{3}=\frac{8}{3}$$

Repeated Addition = Multiplication

$$1\frac{1}{4}=\frac{4}{4}+\frac{1}{4}=\frac{5}{4} \quad \rightarrow \quad 1\frac{1}{4}=\frac{(1\times4)+1}{4}=\frac{5}{4}$$

Step 2: Multiply the improper fractions

$$\frac{8}{3}\times\frac{5}{4}=\frac{\overset{10}{\cancel{40}}}{\underset{3}{\cancel{12}}}=\frac{10}{3}=3\frac{1}{3}$$

HELPFUL HINT

Remember to simplify when possible. Look for common factors
$3 \times 4 = 12$ *and* $10 \times 4 = 40$.

Step 3: State your answer

$$2\frac{2}{3}\text{ feet}\times1\frac{1}{4}\text{ feet}=3\frac{1}{3}\text{ square feet}$$

You need $3\frac{1}{3}$ square feet of tile for the wall behind the sink.

PRACTICE—CHECK FOR UNDERSTANDING: Multiplying Mixed Numbers

1. Mai used $2\frac{1}{4}$ cups of flour to make 1 cake. How many cups of flour are needed to make 4 cakes? Show your work in the box below.

2. The local dry cleaners mailed out the flyer below. Your mom is going to take advantage of the special.

Snow Day Special

Drop off in the AM (by 10AM)

pick up same day in the PM

Pay by the pound

1 pound of clothes $3

What will she pay for $7\frac{2}{3}$ pounds of clothes? Show your work in the box below.

3. The office sitting area is getting new carpet. Your math teacher asked the class to calculate how many square feet of carpet will have to be purchased. The sitting area has the dimensions of $8\frac{1}{4}$ feet × $9\frac{1}{3}$ feet. What is the total square footage of carpet needed for the sitting area?

4. The flight from Newark Airport to San Juan, Puerto Rico is $3\frac{3}{4}$ hours long. It takes $5\frac{1}{3}$ times as long to get to Sydney, Australia from Newark. How long does it take to get to Sydney, Australia? Show your work in the box below.

Answers (pages 161–162)

Question	Answer	Detailed Explanation
1	9	$\frac{9}{4}\times 4=9$
2	\$23	$\frac{23}{3}\times 3=23$
3	77 square feet	$\frac{33}{4}\times\frac{28}{3}=\frac{11\times 3}{4}\times\frac{7\times 4}{3}=11\times 7=77$
4	20 hours	$\frac{15}{4}\times\frac{16}{3}=5\times 4=20$

Using Fractions to Scale (Resize)

THE STANDARD

CCSS.5.NF.5 Interpret multiplication as scaling (resizing), by

a. Comparing the size of a product to the size of one factor on the basis of the size of the other factor, without performing the indicated multiplication.

b. Explaining why multiplying a given number by a fraction greater than 1 results in a product greater than the given number (recognizing multiplication by whole numbers greater than 1 as a familiar case); explaining why multiplying a given number by a fraction less than 1 results in a product smaller than the given number; and relating the principle of fraction equivalence $\frac{a}{b}=\frac{(n\times a)}{(n\times b)}$ to the effect of multiplying $\frac{a}{b}$ by 1.

What Does This Mean?

What does scaling mean? Have you ever heard the saying "I need to scale back"? What would you be doing if you scaled back? In reality, it means you would be doing less of something. How about using the saying, "Double the fun"? That would mean two times the regular fun. So, words like quadruple, triple, half, double, and one quarter mean that you are changing the size or quantity of an object. You are either stretching the object or shrinking the object. More importantly, you are able to compare the size of the object based upon the words used without performing the actual multiplication.

If I were to say that a pothole doubled in size and that another pothole is half the size of the first pothole. I have just presented you with an image of two ways to scale a quantity without multiplying.

In this section, you will learn how to look at and reason about mathematics. You will begin to see scaling as an if–then situation. If you multiply a number by a fraction less than 1, then what is the result of the product? Does it stretch or shrink? Let's move forward and see.

When you think about scaling there are three key points for you to remember. Instead of telling you what the three key points are, I want you to think and reason about the three scenarios below.

Scenario 1

Asia ran 8 miles in 1 hour on Monday. Today she only had time to run for $\frac{1}{2}$ an hour.

Did Asia run more or less than 8 miles today?

Think about what is happening in this scenario. Is she running an equal amount of time, more time, or less time than Monday?

Example 1

$$\text{If } \left(\frac{a}{b}\right) < 1, \text{ then } \left(\frac{a}{b}\right) \times q < q \text{ or } \left(\frac{a}{b}\right) \times \left(\frac{c}{d}\right) < \frac{c}{d}.$$

Looking at Scenario 1, $\left(\frac{1}{2}\right) < 1$, so $\left(\frac{1}{2}\right) \times 8 = \frac{1 \times 8}{2} = 4$

KEY CONCEPT

If you multiply a number by a fraction less than 1, the product will be less than the number.

Scenario 2

Asia ran 8 miles in 1 hour on Monday. Today she had some additional time so she ran for $1\frac{3}{4}$ hours. Did she run more than, less than, or the same 8 miles today?

In this scenario the amount of time she spends running is more than Monday.

Example 2

$$\text{If } \left(\frac{a}{b}\right) > 1, \text{ then } \left(\frac{a}{b}\right) \times q > q \text{ or } \left(\frac{a}{b}\right) \times \left(\frac{c}{d}\right) > \frac{c}{d}.$$

Looking at Scenario 2, $\left(1\frac{3}{4}\right) > 1$, so $\left(1\frac{3}{4}\right) \times 8 = \frac{7}{4} \times 8 = \frac{7 \times \cancel{8}}{\cancel{4}} = \frac{7 \times 8}{4} = \frac{7 \times 2}{1} = 14$

KEY CONCEPT

If you multiply a number by a fraction greater than 1, the product will be greater than the number.

Scenario 3

Asia ran 8 miles in 1 hour on Monday. Today she had exactly the same amount of time to run as she did on Monday so she ran for $\frac{4}{4}$ of an hour. Did she run more than, less than, or the same 8 miles today?

In this scenario the amount of time she spends running is equal to Monday.

Example 3

$$\text{If } \left(\frac{a}{b}\right) = 1, \text{ then } \left(\frac{a}{b}\right) \times q = q \text{ or } \left(\frac{a}{b}\right) \times \left(\frac{c}{d}\right) = \frac{c}{d}.$$

Looking at Scenario 3, $\left(\frac{4}{4}\right) = 1$, so $\left(\frac{4}{4}\right) \times 8 = \frac{4 \times 8}{4} = 8$

KEY CONCEPT

If you multiply a number by a fraction equal to 1, the product will be equal to the number.

PRACTICE—CHECK FOR UNDERSTANDING: Using Fractions to Scale

1. A number is multiplied by the fraction $\frac{2}{3}$, which of the following is true?

○ A. The product is equal to the $\frac{2}{3}$.

○ B. The product is less than the number.

○ C. The product is greater than the number.

○ D. The product is greater than $\frac{2}{3}$.

2. A number is multiplied by the fraction $\frac{7}{4}$, which of the following is always true?

○ A. The product is greater than $\frac{7}{4}$.

○ B. The product is less than $\frac{7}{4}$.

○ C. The product is greater than the number.

○ D. The product is less than the number.

3. Select the **three** inequalities that are true.

□ A. $\frac{5}{6}\times\frac{3}{8}<\frac{3}{8}$

□ B. $7\times2\frac{3}{4}<7$

□ C. $\frac{2}{5}\times\frac{2}{3}<\frac{2}{3}$

□ D. $\frac{11}{8}\times4>4$

4. You are making popcorn for movie night at home. The recipe says use 1 cup of popcorn to make 8 servings. You use only $\frac{3}{4}$ cup of popcorn. How many servings of popcorn will you make?

○ A. Greater than 8

○ B. Exactly 8

○ C. Less than 8

○ D. 0

5. Your school's state test scores were $5\frac{3}{10}$ times higher this year (2014 results) than in 2013. Circle all answers that apply.

Scores were:

☐ A. Higher this year ☐ B. Higher in 2013

☐ C. The same as last year ☐ D. Lower last year

6. The group was given the following math problem for group work.

Mr. Oluwole commutes to and from work Monday through Friday. He uses $\frac{7}{8}$ gallons of gas to drive to and from work daily. Will he use more or less than 5 gallons of gas in the week?

Jack, a member of your group, wrote that Mr. Oluwole will use more than 5 gallons of gas during the week.

- Without solving the problem, use the space below to explain why Jack's answer is incorrect.

- Use your understanding of scaling to determine how much gas Mr. Oluwole uses in 5 days.

Answers (pages 166–167)

Question	Answer	Detailed Explanation
1	B	
2	C	
3	A C D	A and C: Both fractions are being multiplied by a fraction less than 1. As such the product will be less than the original numbers $\frac{3}{8}$ and $\frac{2}{3}$. D: $\frac{11}{8} > 1$. The product has to be greater than 4.
4	C	The amount of popcorn you use is less than 1.
5	A, D	
6	See Detailed Explanation.	Jack did not consider that $\frac{7}{8}$ is less than 1. Since $\frac{7}{8} < 1$, then $\frac{7}{8} \times 5 < 5$. Mr. Oluwole uses $\frac{7}{8} \times 5 = \frac{35}{8} = 4\frac{3}{8}$ gallons of gas

Comprehending Fractions as Division

THE STANDARD

CCSS.5.NF.3. Interpret a fraction as division of the numerator by the denominator $\left(\frac{a}{b} = a \div b\right)$. Solve word problems involving division of whole numbers leading to answers in the form of fractions or mixed numbers (e.g., by using visual fraction models or equations to represent the problem). For example, interpret $\frac{3}{4}$ as the result of dividing 3 by 4, noting that $\frac{3}{4}$ multiplied by 4 equals 3, and that when 3 wholes are shared equally among 4 people each person has a share of size $\frac{3}{4}$. If 9 people want to share a 50-pound sack of rice equally by weight, how many pounds of rice should each person get? Between what two whole numbers does your answer lie?

What Does This Mean?

If you have 1 candy bar and you share it **equally** with a friend. Each of you now has $\frac{1}{2}$ of the candy bar. Well, in that scenario, you just divided the candy bar into 2 **equal** pieces. So when you think about fractions think about the process of equally dividing a whole number (the numerator) by another whole number (the denominator). Let's continue to look at how fractions can be thought of as division.

Fractions Are Division?

Three friends want to equally share 2 hot dogs. How much of each hot dog does each friend receive?

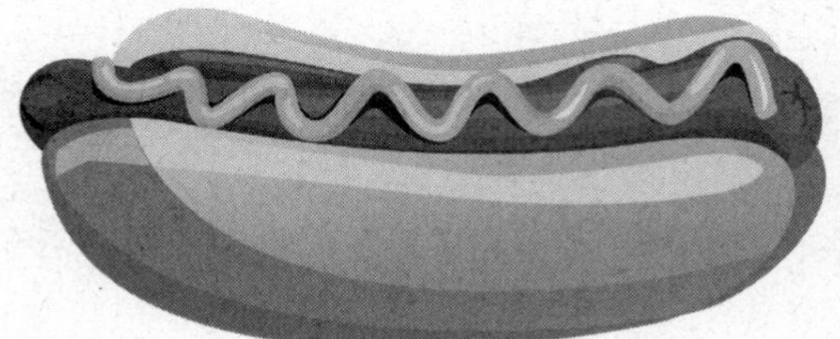
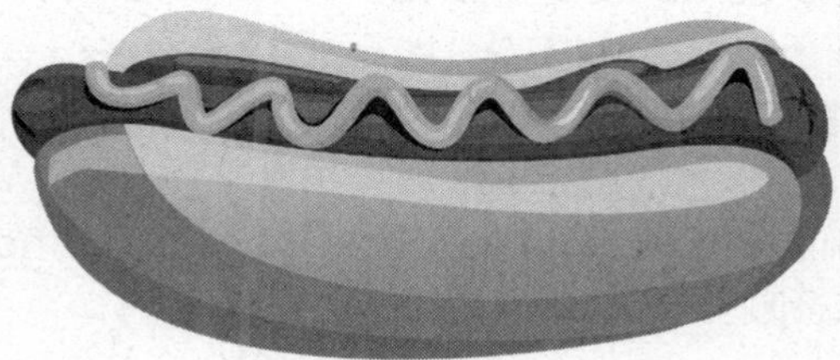

Think about this. We have 2 hot dogs and 3 people.

We have to make sure that each person has an equal share. So you need to partition each hot dog into 3 equal parts.

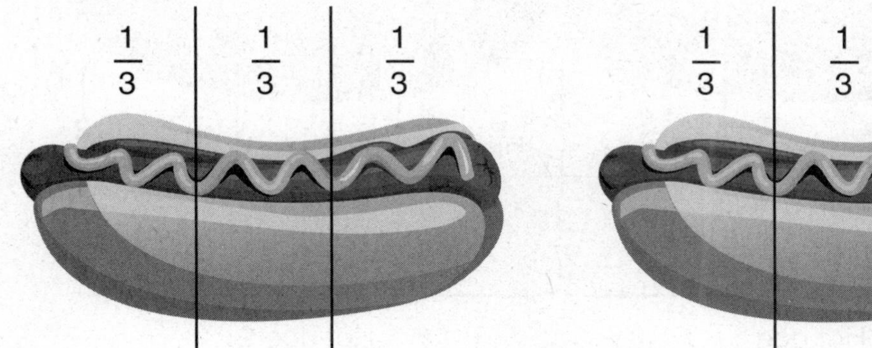

Friend 1	Friend 2	Friend 3
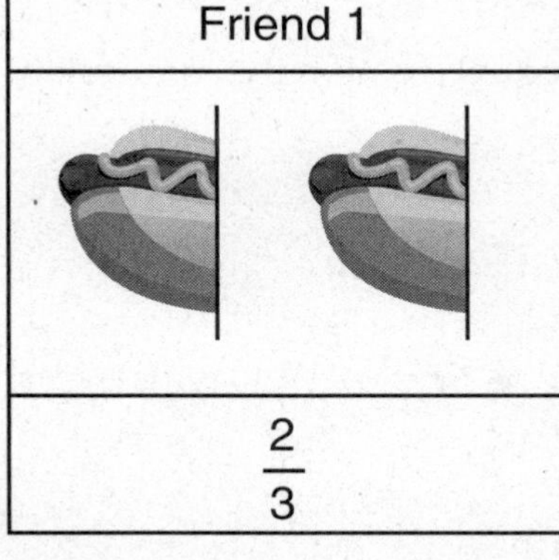	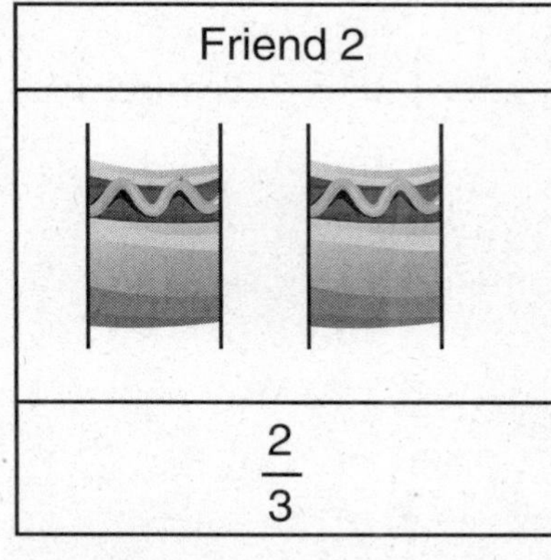	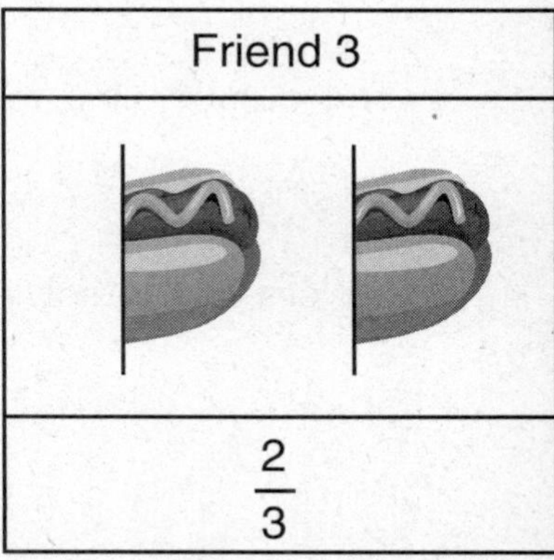
$\frac{2}{3}$	$\frac{2}{3}$	$\frac{2}{3}$

So you can see that 2 hot dogs divided equally among 3 people can be represented as

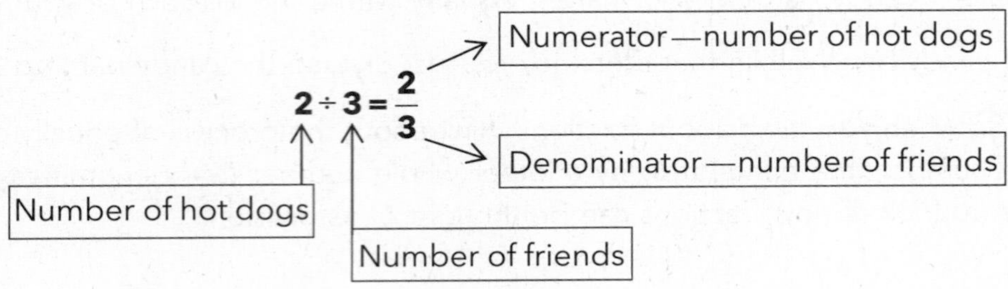

Fractions Are Division? Modeling Using the Number Line

You can model the same hot dog problem using a number line.

Problem

Three friends want to equally share 2 hot dogs. How much of each hot dog does each friend receive?

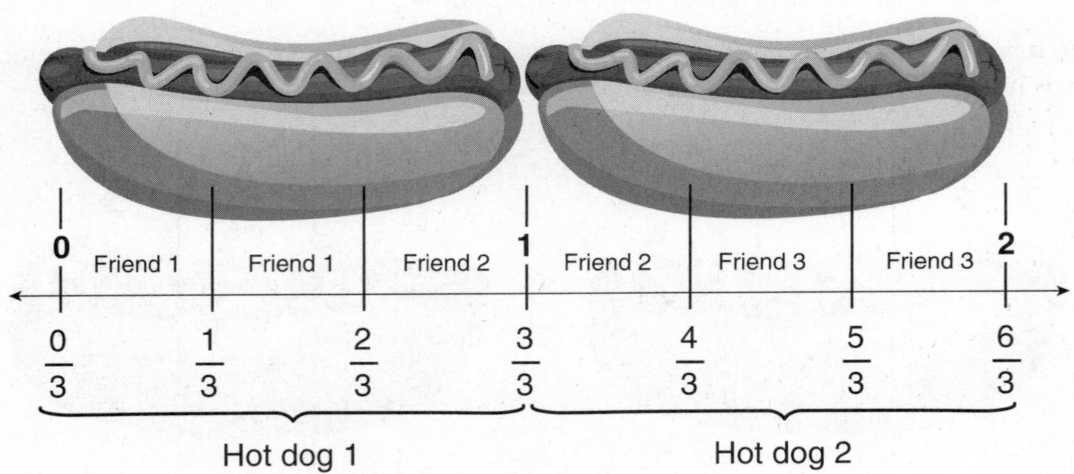

The number line model shows that 2 divided into 3 equal sections is equivalent to $\frac{2}{3}$.

So let's look at solving some problems.

PRACTICE—CHECK FOR UNDERSTANDING: Fractions as Division

1. Circle the correct fraction to describe the scenarios listed in the table below.

Scenario	Fraction	
Eight friends share 5 pizzas equally.	$\frac{8}{5}$	$\frac{5}{8}$
Fifteen friends equally share 5 packs of crayons.	$\frac{1}{3}$	$\frac{5}{15}$
Twenty pounds of flour distributed equally among 12 bags to be shipped out to the local bakeries.	$\frac{12}{20}$	$1\frac{2}{3}$
Ten students equally share 7 yards of fabric in sewing class.	$\frac{10}{7}$	$\frac{7}{10}$

2. Draw a line to connect each fraction or mixed number to the division expression that it represents. All expressions may or may not be used.

Fraction or Mixed Number	Division Expression
$\frac{2}{3}$	$11 \div 3$
	$11 \div 6$
$\frac{3}{11}$	$2 \div 3$
	$6 \div 5$
$1\frac{5}{6}$	$3 \div 11$

3. Five friends shared 2 packs of baseball cards. How much of each pack did each friend get? Place your answer in the box below.

4. The teacher gave you 55 minutes to complete 35 math questions. How much time do you have to complete each problem? Place your answer in the box below.

5. Celestina needs 9 equal sections of construction paper from a roll that is 15 meters long. Select yes or no for each expression that correctly represents 1 section of the construction paper.

A. $\frac{15}{9}$	YES	NO
B. $\frac{3}{5}$	YES	NO
C. $1\frac{2}{3}$	YES	NO
D. $15 \div 9$	YES	NO
E. $9 \div 15$	YES	NO

Answers (pages 171–172)

<table>
<tr><th>Question</th><th>Answer</th><th>Detailed Explanation</th></tr>
<tr><td>1</td><td></td><td>
<table>
<tr><th>Scenario</th><th colspan="2">Fraction</th></tr>
<tr><td>Eight friends share 5 pizzas equally.</td><td>$\frac{8}{5}$</td><td>$\frac{5}{8}$</td></tr>
<tr><td>Fifteen friends equally share 5 packs of crayons.</td><td>$\frac{1}{3}$</td><td>$\frac{5}{15}$</td></tr>
<tr><td>Twenty pounds of flour distributed equally among 12 bags to be shipped out to the local bakeries.</td><td>$\frac{12}{20}$</td><td>$1\frac{2}{3}$</td></tr>
<tr><td>Ten students equally share 7 yards of fabric in sewing class.</td><td>$\frac{10}{7}$</td><td>$\frac{7}{10}$</td></tr>
</table>
</td></tr>
<tr><td>2</td><td></td><td>
<table>
<tr><th>Fraction or Mixed Number</th><th>Division Expression</th></tr>
<tr><td></td><td>$11 \div 3$</td></tr>
<tr><td>$\frac{2}{3}$</td><td></td></tr>
<tr><td></td><td>$11 \div 6$</td></tr>
<tr><td>$\frac{3}{11}$</td><td>$2 \div 3$</td></tr>
<tr><td></td><td>$6 \div 5$</td></tr>
<tr><td>$1\frac{5}{6}$</td><td></td></tr>
<tr><td></td><td>$3 \div 11$</td></tr>
</table>
</td></tr>
<tr><td>3</td><td>$\frac{2}{5}$ pack</td><td></td></tr>
<tr><td>4</td><td>$\frac{55}{35} = \frac{11}{7} = 1\frac{4}{7}$ minutes</td><td></td></tr>
<tr><td>5</td><td>A. yes
B. no
C. yes
D. yes
E. no</td><td></td></tr>
</table>

Dividing Fractions and Whole Numbers

THE STANDARD

CCSS.NF.7. Apply and extend previous understandings of division to divide unit fractions by whole numbers and whole numbers by unit fractions.[2]

A. Interpret division of a unit fraction by a nonzero whole number, and compute such quotients. For example, create a story context for $\frac{1}{3} \div 4$, and use a visual fraction model to show the quotient. Use the relationship between multiplication and division to explain that $\frac{1}{3} \div 4 = \frac{1}{12}$ because $\frac{1}{12} \times 4 = \frac{1}{3}$.

B. Interpret division of a whole number by a unit fraction, and compute such quotients. For example, create a story context for $4 \div \frac{1}{5}$, and use a visual fraction model to show the quotient. Use the relationship between multiplication and division to explain that $4 \div \frac{1}{5} = 20$ because $20 \times \frac{1}{5} = 4$.

C. Solve real-world problems involving division of unit fractions by nonzero whole numbers and division of whole numbers by unit fractions (e.g., by using visual fraction models and equations to represent the problem). For example, how much chocolate will each person get if 3 people share $\frac{1}{2}$ lb of chocolate equally? How many $\frac{1}{3}$-cup servings are in 2 cups of raisins?

[2] Students able to multiply fractions in general can develop strategies to divide fractions, by reasoning about the relationship between multiplication and division. But division of a fraction by a fraction is not a requirement at this grade.

What Does This Mean?

The best way to explain dividing fractions is by using a rectangular area model. Remember that multiplication is the opposite operation of division. Take for example the multiplication problem using whole numbers: $4 \times 2 = 8$; relating multiplication to division you know that $8 \div 4 = 2$. However, dividing $8 \div 4$ will give you the same result as multiplying $8 \times \frac{1}{4}$.

×	2	
4	1	2
	3	4
	5	6
	7	8

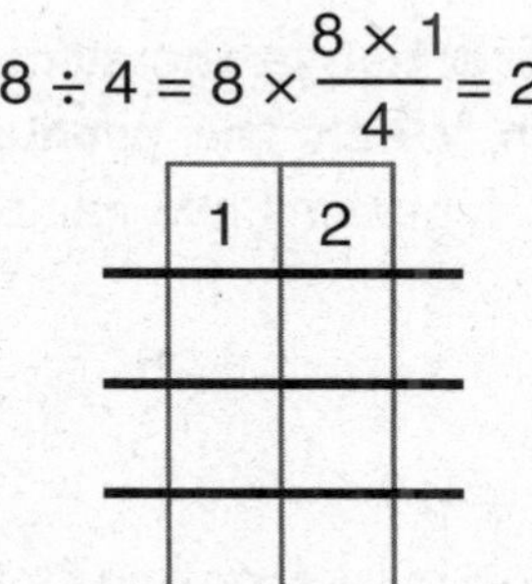

HELPFUL HINT

When you are dividing a number by a unit fraction (the numerator is always 1), you can also solve fraction-division with multiplication.

What Happens When You Divide a Unit Fraction by a Whole Number?

Imagine dividing $\frac{1}{5}$ pound of candy among 3 friends. That means you need to divide $\frac{1}{5}$ into 3 equal parts. Let's use the rectangular area model below to represent the problem.

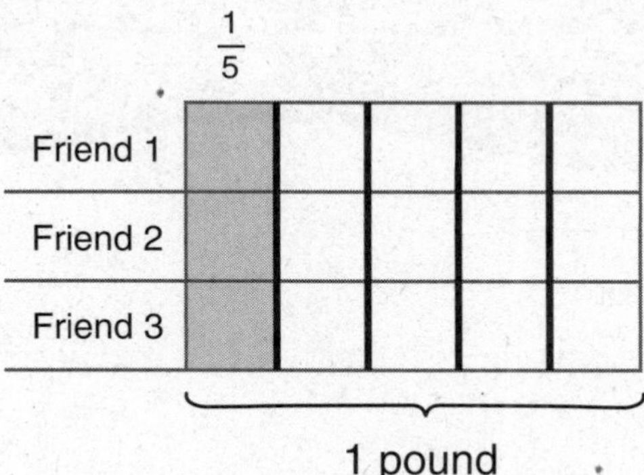

In other words, $\frac{1}{5} \div 3$ means that if you divide $\frac{1}{5}$ pound of candy into 3 equal parts, each friend receives $\frac{1}{3}$ of the $\frac{1}{5}$ pound of candy or $\frac{1}{15}$.

$$\frac{1}{3} \times \frac{1}{5} = \frac{1}{15}$$

KEY CONCEPT

Dividing by a fraction is the same as multiplying by the **reciprocal** of the fraction. **When the product of two fractions is equivalent to 1**, the fractions are reciprocals.

$$3 \times \frac{1}{3} = \frac{3}{1} \times \frac{1}{3} = \frac{3}{3} = 1$$

So note, division by a fraction is the same as multiplying by the reciprocal of the fraction.

What Happens When You Divide a Whole Number by a Unit Fraction?

Imagine slicing a 3 foot sub into $\frac{1}{5}$ foot sections. How many $\frac{1}{5}$ portions are there in 3? Let's use a number line to model this situation.

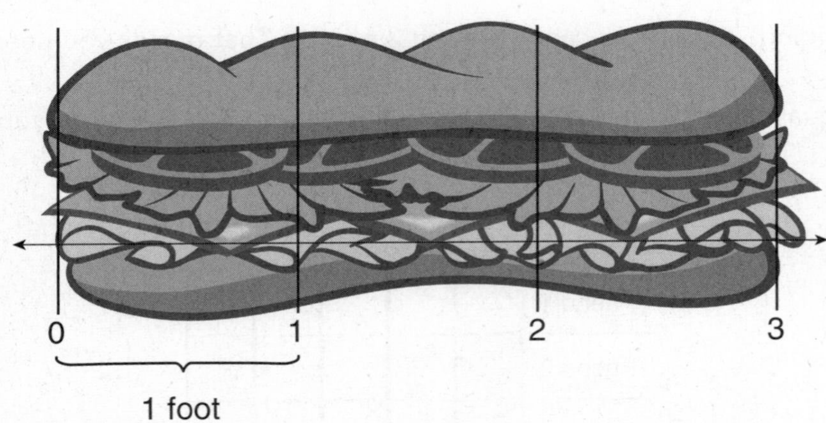

You can see from the $\frac{1}{5}$ markings on the number line that there are fifteen $\frac{1}{5}$ foot sections in 3 feet.

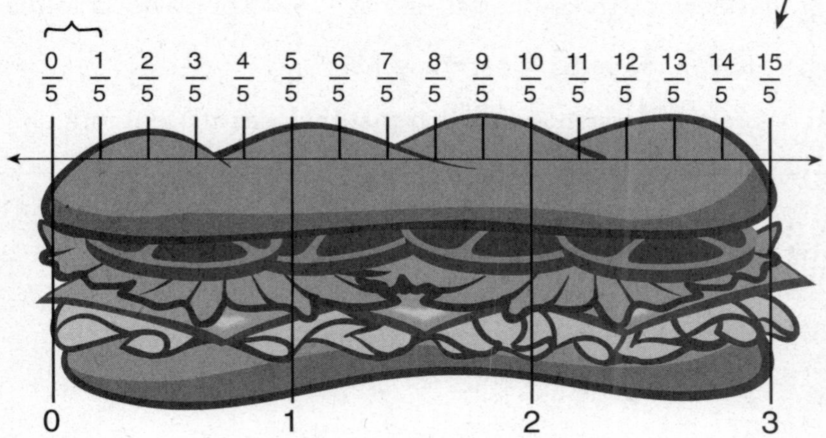

In other words, $3 \div \frac{1}{5}$, means if you divide a 3 foot sub into $\frac{1}{5}$ equal parts, you will have 15 parts.

$$3 \times 5 = 15$$

HELPFUL HINT

The reciprocal

$$\frac{1}{5} \times \frac{5}{1} = \frac{1}{5} \times 5 = 1$$

PRACTICE—CHECK FOR UNDERSTANDING: Dividing Fractions and Whole Numbers

1. Salvador makes wooden keychains. He has $\frac{1}{4}$ yard of plywood to make 3 keychains. How many yards is each keychain?

Part A: Write the division expression that matches the description in the story.

Part B: Use the space below to create an area model to determine how many yards each keychain is.

2. Create a number line model for the story below.

Your class ordered a 12-foot sub for the President's Day Concert. The teacher cut it into $\frac{1}{3}$ equal parts to make sure there was enough to serve everyone in the band. How many pieces were needed to serve the members in the band?

3. Your mother bought $\frac{1}{5}$ pound of lunch meat to share equally among 3 sandwiches. How many pounds of meat are on each sandwich? Draw a rectangular area model to represent this problem.

4. $8 \div \frac{1}{3} =$ ____________

5. $4\frac{1}{5} \div 4 =$ ____________

6. $\frac{1}{4} \div 3 =$ ____________

7. $\frac{1}{8} \div \frac{1}{6} =$ ____________

8. $12 \div \frac{1}{10} =$ ____________

9. $3\frac{1}{10} \div \frac{1}{3} =$ ____________

Answers (pages 178–180)

Question	Answer	Detailed Explanation
1	Part A: $\frac{1}{4} \div 3$ Part B: See Detailed Explanation. Each keychain is $\frac{1}{12}$ yard.	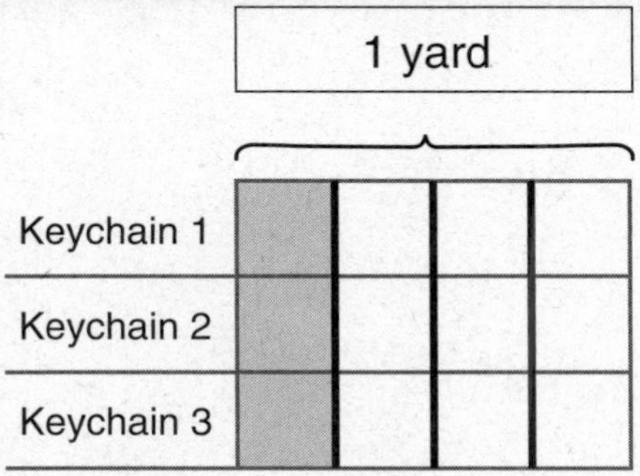 1 yard Keychain 1 Keychain 2 Keychain 3
2	Any number line model that shows the sandwich divided into $\frac{36}{3}$	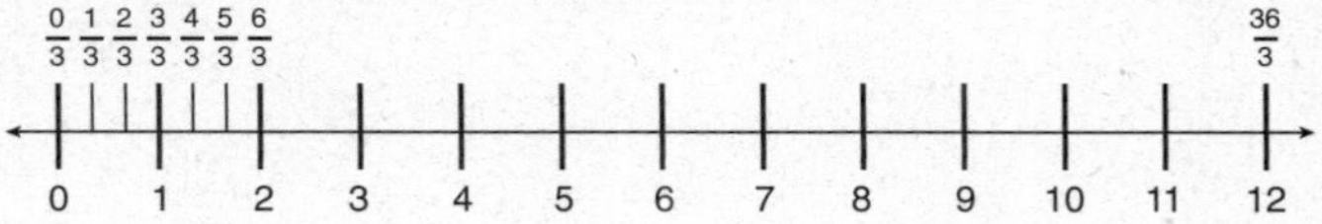 $\frac{0}{3}$ $\frac{1}{3}$ $\frac{2}{3}$ $\frac{3}{3}$ $\frac{4}{3}$ $\frac{5}{3}$ $\frac{6}{3}$ … $\frac{36}{3}$ 0 1 2 3 4 5 6 7 8 9 10 11 12
3	Each sandwich has $\frac{1}{15}$ pound of lunch meat.	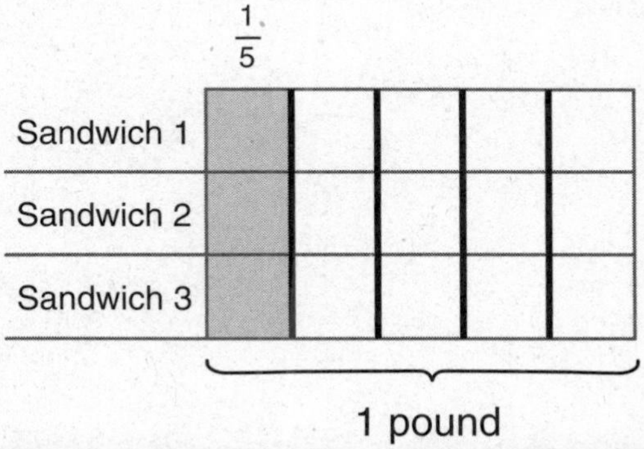 $\frac{1}{5}$ Sandwich 1 Sandwich 2 Sandwich 3 1 pound

Question	Answer	Question	Answer
4	24	5	$\frac{21}{20} = 1\frac{1}{20}$
6	$\frac{1}{12}$	7	$\frac{6}{8} = \frac{3}{4}$
8	120	9	$\frac{93}{10} = 9\frac{3}{10}$

Measurement and Data

CHAPTER 6

Have you ever heard the saying give a person an inch and they will take a yard? Maybe you have, maybe not. The importance of using this example is not what the saying implies but more so what is implied by the math. An inch is a small unit of measurement compared to a yard. By the way, the saying implies that a person will take advantage of a given opportunity. For example, "Mom, can I go across the street to my friend's house for 1 hour?" Your mom agrees, and then you stay gone for 3 hours. Sure you know what it means now! However, think about the math in the saying give a person an inch and they will take a yard! How many inches are in a yard and how many yards are in an inch? When you move back and forth across standard units of measurement, you are said to be converting the measurements. Possibly you never stopped to think about the relationship that exists between the units of measurement. What's the connection between pounds and ounces, inches and feet, centimeters and meters? Hopefully, I have you wondering, and possibly you will think about the math as you try to figure out the next saying.

What does the saying "an inch deep and mile wide" mean? Again I am really not as focused on the meaning of the saying as much as I am the math. The saying actually refers to someone who claims to have a little knowledge about everything but nothing in depth. For example, I am only a football fan come the playoffs and the Super Bowl. My team lost the Super Bowl this year due to a crazy play at the end zone! They were on the 1-yard line and threw a pass that was intercepted! Anyway, if you asked me anything about the team's record for the season I couldn't tell you. But I know enough about the calls and the plays to have a conversation about football while I'm watching a game. So let's go back to thinking about the math. How deep is something if it only measures an inch? Compare a pool that is 1-inch deep to a pool 3-feet deep. The key here is that when you start looking at the "depth" of something you are talking about having an understanding of volume. A soda bottle has volume, a box has volume. What else can you think of that has volume?

You worked with measurement in the earlier grades. Now let's focus on converting between units and understanding what volume measures.

Measurement Conversions

THE STANDARD

CCSS.5.MD.1. Convert among different-sized standard measurement units within a given measurement system (e.g., convert 5 cm to 0.05 m), and use these conversions in solving multi-step, real-world problems.

What Does This Mean?

Before we can talk about measurement units, let's identify what we measure:

Length, Capacity (how much something holds or contains),
Weight, Mass, and Time

HELPFUL HINT

Weight and mass are used interchangeably; however, they do not exactly measure the same thing. Mass measures what an object is composed of; weight measures the force on the object due to gravity. Don't panic, you will learn more about this in science. For now, just think of them as the same for the purpose of measuring objects.

The table below shows the abbreviations for Customary and Metric Units.

Measurement Type	Customary Units	Metric Units
Length	Inch (in) Foot (ft) Yard (yd) Mile (mi)	Millimeter (mm) Centimeter (cm) **Meter (m)** Kilometer (km)
Capacity	Cup (c) Fluid ounce (fl oz) Pint (pt) Quart (qt) Gallon (gal)	Mililiter (ml) **Liter (L)** Cubic centimeter (cc)
Weight/ Mass	Ounce (oz) Pound (lb) Ton (T)	**Gram (g)** Kilogram (kg)
Time	Second (sec) Minute (min) Hour (hr) Day Week (wk) Month (mo) Year (yr)	
		BASE UNIT

In the Metric System, each type of measurement uses a base unit. Meter is the base unit for length, liter is the base unit for capacity, and gram is the base unit for mass.

Identifying Units of Measurement

The customary system of measurement is most commonly used here in the United States and is known as the standard unit of measurement. It is important for you to reason and attend to precision in identifying and selecting the most appropriate customary unit of measurement when measuring things.

For example, what would you choose as the most appropriate unit to measure:

1. The depth of a pool	○ A. inch ○ B. mile ○ C. feet ○ D. yard
2. The weight of a baby	○ A. ounce ○ B. pound ○ C. ton
3. The amount of water in a fish tank	○ A. cup ○ B. quart ○ C. gallon ○ D. fluid ounce

Answers (page 186)

1. *C*; 2. *B*; 3. *C*

1. **Measuring the depth of a pool**. Use the process of elimination. An inch is a very small measurement. By the time you stick your big toe in the pool to test the water temperature, you will have basically measured an inch. An inch is not the answer. A mile? Well, about 20 New York City blocks are considered a mile. If you use a mile to measure the pool, that must be a really deep pool. A mile is not the answer. Feet? Well, this makes sense. If you are 5 feet 3 inches tall, you already know that the water will be over your head if you enter the 6 foot section of a pool. A yard is not the most practical unit of measurement to use either. One yard equals 3 feet. So the 4 foot marker of a pool would read $1\frac{1}{3}$ yards.

2. **Measuring the weight of a baby.** If you have no idea about the size of an ounce, the next time you go to the grocery store make sure to look at some cans of tuna or tomato paste. An ounce is a very small measurement. In fact, a nutritionist suggests that an ounce is about the size of your entire thumb and fatty part of your palm. Well, a baby is definitely bigger than an ounce.

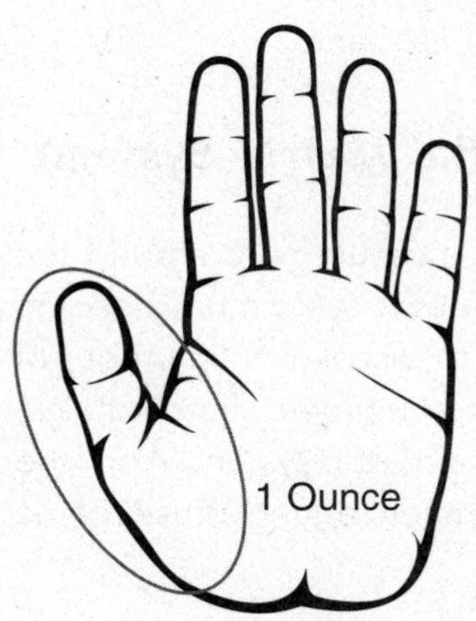

A pound? Well, 16 ounces is 1 pound. A professional football weighs about 15 ounces, which is almost a pound. Now, 2,000 pounds equals 1 ton. That's about the weight of a car, named Herbie, that was really popular when I was a kid. Okay, a baby is not a ton so a pound is the best unit of measurement.

3. **The amount of water in a fish tank**. I'm sure you know a cup of water is not going to fill a fish tank. Four cups equals 1 quart. However, it is still not enough water for a fish tank. Maybe it is enough for a small fish bowl, but we are talking about a fish tank.

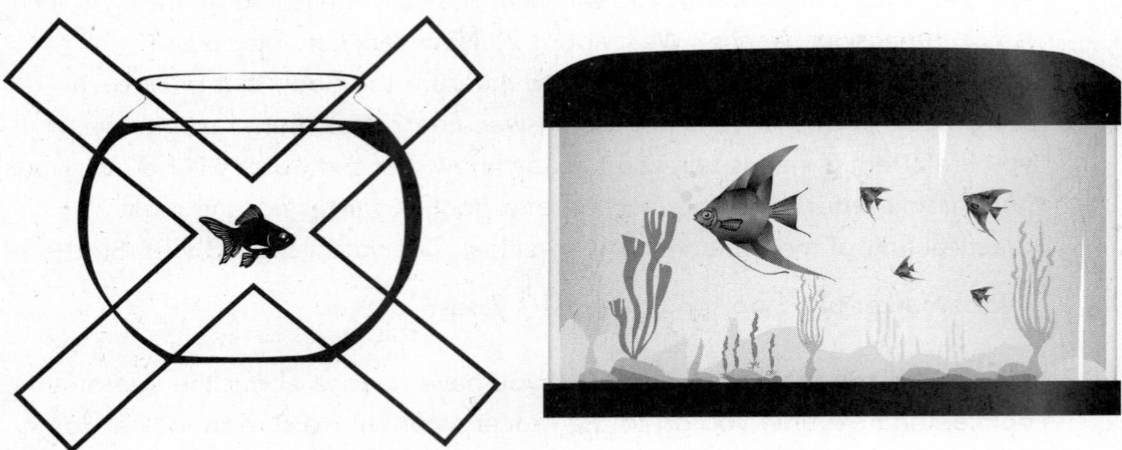

A gallon? Well, there are 4 quarts in a gallon, and it takes 128 fluid ounces to make a gallon.

Obviously, a gallon is the best unit of measurement for the fish tank.

The International System of Measurement (the Metric System)

The Metric System is the most commonly used system of measurement around the world. However, you should also know that the use of the metric system has become widespread throughout our country. You may be surprised to discover the number of items being produced in metric units. In fact, photographic equipment, automobiles, computers, pharmaceutical products, and soft drinks use the metric system. Also, the scientific and medical communities along with certain government agencies use metric units exclusively.

Just as with customary units, it is important for you to reason and attend to precision in identifying and selecting the most appropriate metric unit of measurement when measuring things, too. Since you may not be as familiar with using the metric system, let's try to visualize how small or large metric units of measurement may be.

Measurement Type	Metric Units	
Length	Millimeter (mm) Centimeter (cm) **Meter (m)** Kilometer (km)	The thickness of a video game disc. The diameter of a shirt button. The height the door knob from the floor. Eighty buses lined up front end to back end. (That's a lot of buses.)
Capacity	Milliliter (ml) **Liter (L)** Cubic centimeter (cc or ccm)	Twenty drops of medicine. A regular bottle of soda that serves 5 people. A small sugar cube.
Weight/Mass	**Gram (g)** Kilogram (kg)	The weight of a paper clip. A small cantaloupe.

For example, what would you choose as the most appropriate unit to measure:

1. A marble	○ A. gram ○ B. kilogram ○ C. centimeter ○ D. liter
2. The amount of water in the bathroom sink	○ A. kilogram ○ B. milliliter ○ C. liter
3. The length of an ant	○ A. centimeter ○ B. gram ○ C. meter ○ D. liter

Answers (page 189)

1. *A*; 2. *C*; 3. *A*

1. **A marble.** You are measuring weight or mass. A marble is much smaller than a cantaloupe, which weighs about a kilogram. A kilogram is not a good choice. A centimeter measures length, and a liter measures capacity. The best answer is A, gram.
2. **The amount of water in the bathroom sink.** You are measuring capacity. A milliliter is small and a kilogram measures weight/mass. The best answer is C, liter.
3. **The length of an ant.** You are measuring length. A meter is too long, a gram measures weight/mass, and a liter measures capacity. The best answer is A, centimeter.

You just compared the units of measurement to determine the most appropriate unit of measurement. In this section you will actually convert the measurements to the equivalent measures. The tables on page 191 list the basic conversion factors for the customary and metric units. You are expected to know the shaded units and you should become familiar with the other units.

Also don't forget 1 week = 7 days and 1 year = 12 months.

Metric Units		
Length	**Capacity**	**Weight/Mass**
1 m = 100 cm	1 L = 1,000 mL	1 kg = 1,000 g
1 m = 1,000 mm	1 L = 1,000 cc	
1 km = 1,000 m		

The focus of this chapter is on learning to convert between units of measurement. The most important thing to remember is that to convert from one unit to another you will have to either multiply or divide.

KEY CONCEPT

Customary Units To convert from a larger unit to a smaller unit, multiply or divide.

Metric Units To convert from a larger unit to a smaller unit, you multiply or divide by a power of 10.

Customary Units			
Length	**Capacity**	**Time**	**Weight**
1 ft = 12 in	1 cup = 8 fl oz	1 min = 60 sec	1 lb = 16 oz
1 yd = 36 in	1 pt = 2 cups	1 hr = 60 min	1 T = 2,000 lb
1 yd = 3 ft	1 qt = 2 pt	1 day = 24 hr	
1 mi = 5,280 ft	1 gal = 4 qt		
1 mi = 1,760 yd			

Example Customary Units: Larger unit to a smaller unit—**Multiply**

How many inches are equal to 4 feet?

HELPFUL HINT

Think about the problem in another way. You have 4 feet, how many inches do you have? You are converting from feet (the larger unit) to inches (the smaller unit).

You know 1 foot = 12 inches. To find the number of inches, multiply. 4 feet × 12 = 48 inches.

Example Customary Units: Smaller unit to a larger unit—**Divide**

How many feet are equal to 48 inches?

HELPFUL HINT

Think about the problem in another way. You have 48 inches, how many feet do you have? You are converting from inches (the smaller unit) to feet (the larger unit).

You know 1 foot = 12 inches. To find the number of feet, divide. 48 inches ÷ 12 = 4 feet.

Metric Units		
Length	**Capacity**	**Weight/Mass**
1 m = 100 cm	1 L = 1,000 mL	1 kg = 1,000 g
1 m = 1,000 mm	1 L = 1,000 cc	
1 km = 1,000 m		

Example Metric Units: Larger unit to a smaller unit—**Multiply by a power of 10**

How many centimeters are equal to 3 meters?

HELPFUL HINT

Think about the problem in another way. You have 3 meters, how many centimeters do you have? You are converting from meters (the larger unit) to centimeters (the smaller unit).

You know 1 meter = 100 centimeters. To find the number of centimeters, multiply. 3 meters × 100 = 300 centimeters.

Example Metric Units: Smaller unit to a larger unit—**Divide by a power of 10**

How many meters are equal to 300 centimeters?

HELPFUL HINT

Think about the problem in another way. You have 300 centimeters, how many meters do you have? You are converting from centimeters (the smaller unit) to meters (the larger unit).

You know 1 meter = 100 centimeters. To find the number of meters, divide. 300 centimeters ÷ 100 = 3 meters.

Customary Units			
Length	**Capacity**	**Time**	**Weight**
1 ft = 12 in	1 cup = 8 fl oz	1 min = 60 sec	1 lb = 16 oz
1 yd = 36 in	1 pt = 2 cups	1 hr = 60 min	1 T = 2,000 lb
1 yd = 3 ft	1 qt = 2 pt	1 day = 24 hr	
1 mi = 5,280 ft	1 gal = 4 qt		
1 mi = 1,760 yd			

Metric Units		
Length	**Capacity**	**Weight/Mass**
1 m = 100 cm	1 L = 1,000 mL	1 kg = 1,000 g
1 m = 1,000 mm	1 L = 1,000 cc	
1 km = 1,000 m		

PRACTICE—CHECK FOR UNDERSTANDING: Measurement Conversions

Identify whether the conversion requires you to multiply or divide. Use the tables as your guide.

Conversion Factors	Multiply or Divide	
1. Converting from millimeters to meters.	Multiply	Divide
2. Converting yards to inches.	Multiply	Divide
3. Converting kilograms to grams.	Multiply	Divide
4. Converting from centimeters to meters.	Multiply	Divide

5. Convert the following measurements.

39 yd = ________ft   540 min = ________hr   40 m = ________mm

5 kg = ________g   57 in = ____ft____in   17 min = ________sec

6. Conversions at times can require more than one step.

a. 5 km = ________________cm   b. 10 qt = ________________cups

c. $4\frac{1}{2}$ days = ________________sec   d. 3 mi = ________________in

7. The weather forecaster reported that 2 feet of snow fell in Boston last week. The storm on Monday brought 18 more inches of snow.

- Explain how to find the total snowfall for Boston.
- Find the total snowfall amount for Boston measured in yards.

Show your work in the box below.

8. Game Day Punch calls for 20 fluid ounces of fruit punch, 3 pints of ginger ale, 1 gallon of ice cream, and 32 fluid ounces of mango juice. How many servings of punch are you making if each serving is 1 cup? Show your work in the box below.

Length	Capacity	Time	Weight
1 ft = 12 in	1 cup = 8 fl oz	1 min = 60 sec	1 lb = 16 oz
1 yd = 36 in	1 pt = 2 cups	1 hr = 60 min	1 T = 2,000 lb
1 yd = 3 ft	1 qt = 2 pt	1 day = 24 hr	
1 mi = 5,280 ft	1 gal = 4 qt		
1 mi = 1,760 yd			

Metric Units		
Length	**Capacity**	**Weight/Mass**
1 m = 100 cm	1 L = 1,000 mL	1 kg = 1,000 g
1 m = 1,000 mm	1 L = 1000 cc	
1 km = 1,000 m		

9. Keith bought 5 meters of wood trim. He used 86 centimeters to frame the windows of his dog's house. He used 2 times that amount to make a picture frame for his dad. How much trim, in meters, did he have left after making the picture frame?

- O A. 242 meters
- O B. 2.58 meters
- O C. 2.42 meters
- O D. 258 meters

10. Brown bear cubs weigh about 450 grams at birth. They feed on their mother's milk until the early spring or early summer and weigh about 9 kilograms by then. How much weight, in grams, does a brown bear approximately gain from birth to the summer? Show your work below.

11. Complete the conversion by filling in the box with the correct number below:

30	0.3	3,000	300

3 km = ☐ m

3,000 cm = ☐ m

300 mm = ☐ m

12. Complete the conversion by filling in the box with the correct number below:

48	24	5	2

10 pt = ☐ qt

32 fl oz = ☐ pt

3 gal = ☐ cups

13. How many yards are equivalent to 15,300 feet? Write your answer in the box below.

☐

14. Ahmad is training for track season. He runs 704 yards 3 times a day for 5 days. How many miles does he run in 7 days? Show your work in the box below.

Answers (pages 194–198)

Question	Answer	Detailed Explanation
1	Divide	A millimeter is smaller than a meter.
2	Multiply	A yard is larger than an inch.
3	Multiply	A kilogram is larger than a gram.
4	Divide	A centimeter is smaller than a meter.
5	$39 \times 3 = 117$ ft	1 yd = 3 ft
	$540 \div 60 = 9$ hr	1 hr = 60 min
	$40 \times 1{,}000 = 40{,}000$ mm	1 m = 1,000 mm
	5,000 g	1 kg = 1,000 g
	$57 \div 12 = 4\frac{3}{4}\text{ ft} = 4\text{ ft } 9\text{ in}$ The answer needs to be in inches. What is $\frac{3}{4}$ of 12 inches? Remember how to multiply a fraction and a whole number $\frac{3}{4} \times 12 = \frac{36}{4} = 9$	1 ft = 12 in
	1,020 sec	1 min = 60 sec
6 a–d	A. $5 \times 1{,}000 = 5{,}000$ m $5{,}000 \times 100 =$ 500,000 cm	1 km = 1,000 m You are converting from kilometers to meters (a larger unit to a smaller unit). 1 m= 100 cm Once you know the number of meters you will then need to convert from meters to centimeters (a larger unit to a smaller unit).
	B. $10 \times 2 = 20$ pt $20 \times 2 = 40$ cups	Quart is a larger unit than cup. You can convert to pints to assist you in finding cups. 1 qt = 2 pt 1 pt = 2 cups 2 pt = 4 cups Using your power of 10 understanding 20 pt = 40 cups

Question	Answer	Detailed Explanation
	C. $4 \times 24 = 96 + 12 =$ 108 hours $108 \times 3{,}600 =$ 388,800 seconds	Reason 1 day = 24 hours so $\frac{1}{2}$ = 12 hours 4 days = 96 hours + 12 = 108 hours 1 hr = 60 min and 1 min = 60 sec. There are 3,600 sec in an hour
	D. 3 mi = 190,080 in	Remember 1 mile = 1,760 yards 3 yards = 36 inches You are converting from miles to inches, a larger unit to a smaller unit, so you need to multiply. $3 \times 1{,}760 = 5{,}280$ yd 3 miles is equal to 5,280 yards. You are not at the final answer; you are looking to find inches. Now you need to convert the number of yards to inches. $5{,}280 \times 36 = 190{,}080$ in So 3 miles is the equivalent of 190,080 inches.
7	$2 \times 12 = 24$ inches $24 + 18 = 42$ inches 42 in = 36 in + 6 in = 1 yd 6 in You need the answer in yards. $6\text{ in} \div 36\text{ in} = \frac{6}{36} = \frac{1}{6}$ Final answer $1\frac{1}{6}$ yd	1 ft = 12 in 1 yd = 3 ft 1 yd = 36 in
8	28.5 servings or $28\frac{1}{2}$ servings	20 fl oz = 2.5 cups 3 pt = 6 cups 1 gal = 16 cups 32 fl oz = 4 cups
9	C	5 m = 500 cm $86 \times 2 = 172$ cm + 86 cm = 258 cm 500 cm − 258 cm = 242 cm Convert back to meters $242 \div 100 = 2.42$ m

Question	Answer	Detailed Explanation
10	See Detailed Explanation.	9 kg = 9,000 g 9,000 – 450 = 8,550 g Brown bears gain approximately 8,550 grams from birth to summer.
11	3 km = 3000 m 3000 cm = 30 m 300 mm = 0.3 m	
12	10 pt = 5 qt 32 fl oz = 2 pt 3 gal = 48 cups	
13	See Detailed Explanation.	1 yd = 3 ft 15,300 ÷ 3 = 5,100 yd
14	6 miles	[(704 × 3) × 5] ÷ 1,760 = 6

Interpreting Data

THE STANDARD

CCSS.5.MD.2. Make a line plot to display a data set of measurements in fractions of a unit ($\frac{1}{2}, \frac{1}{4}, \frac{1}{8}$). Use operations on fractions for this grade to solve problems involving information presented in line plots. For example, given different measurements of liquid in identical beakers, find the amount of liquid each beaker would contain if the total amount in all the beakers were redistributed equally.

What Does This Mean?

Imagine that you were tracking the growth of a bean plant in your science class. Once a week your teacher asked you to measure your plant to identify how much it grew and record it in your notebook for 12 weeks. You have these measurements and want to know if there is a better way for you to organize them to make better sense of them?

Well, yes there is! A basic way to organize and display data is by using a line plot. The line plot can assist you in seeing patterns and answering more complex questions about your data.

What Is a Line Plot?

A **line plot** is a graph that shows how often a number occurs. The numbers are called **data**. The data is recorded on a number line by placing an X above the value on the number line. A line plot is a quick, simple way to organize data when you are working with a small set of numbers.

Example: Measuring the Growth of a Bean Plant

You have recorded the following measurements.

$$\frac{1}{2}\text{ in}, \frac{1}{2}\text{ in}, \frac{1}{8}\text{ in}, \frac{1}{4}\text{ in}, \frac{1}{8}\text{ in}, \frac{1}{2}\text{ in}, \frac{1}{8}\text{ in}, \frac{1}{4}\text{ in}, \frac{1}{8}\text{ in}, \frac{1}{4}\text{ in}, \frac{1}{8}\text{ in}, \frac{1}{2}\text{ in}$$

Each piece of numerical information that has been collected is called *data*. The entire collection of your data is considered to be your **data set.** The data set above has 12 entries.

What do the 12 pieces of data mean? Can the data assist you in making any statements about the plant's growth?

Well, let's look at plotting the data on a line plot to determine if you can get a better look at what is happening with the bean plant.

First you need to make a number line.

STEP 1: Let's draw and label a number line that includes your data values.

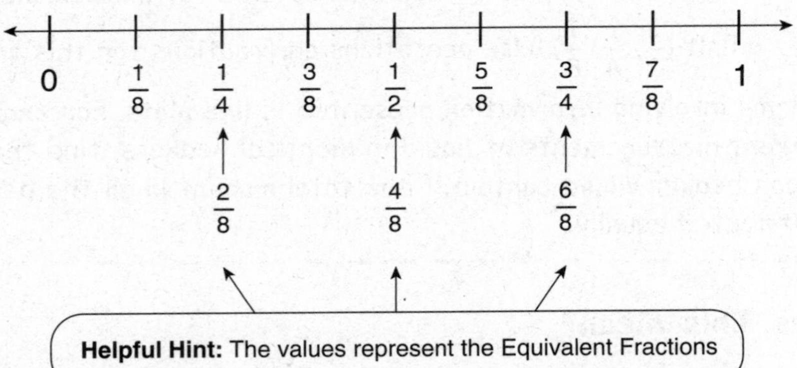

STEP 2: This will be your line plot. Label your line plot "Weekly Growth of Bean Plant (in inches)."

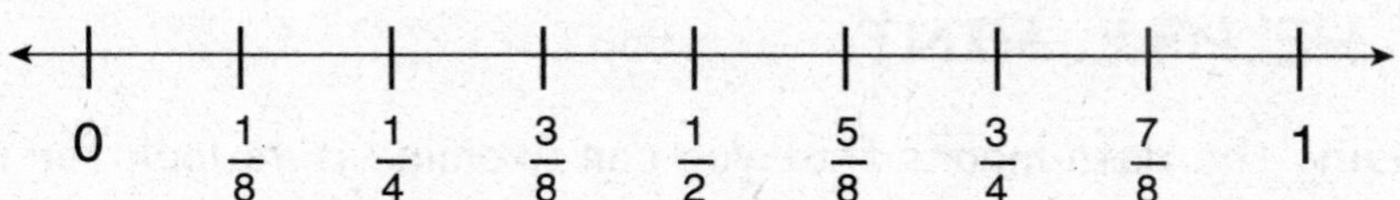

Weekly Growth of Bean Plant (in inches)

STEP 3: Use an X to represent your measurements. Place an X above each number on the number line that matches the data from your data set. Do this until you have recorded all of your data.

HELPFUL HINT

Dots can be used instead of X. This is called a dot plot.

Example

There are 3 measurements equal to $\frac{1}{4}$ in. You should have three X marks above $\frac{1}{4}$.

Data set $\left[\frac{1}{2}\text{ in}, \frac{1}{2}\text{ in}, \frac{1}{8}\text{ in}, \frac{1}{4}\text{ in}, \frac{1}{8}\text{ in}, \frac{1}{2}\text{ in}, \frac{1}{8}\text{ in}, \frac{1}{4}\text{ in}, \frac{1}{8}\text{ in}, \frac{1}{4}\text{ in}, \frac{1}{8}\text{ in}, \frac{1}{2}\text{ in}\right]$

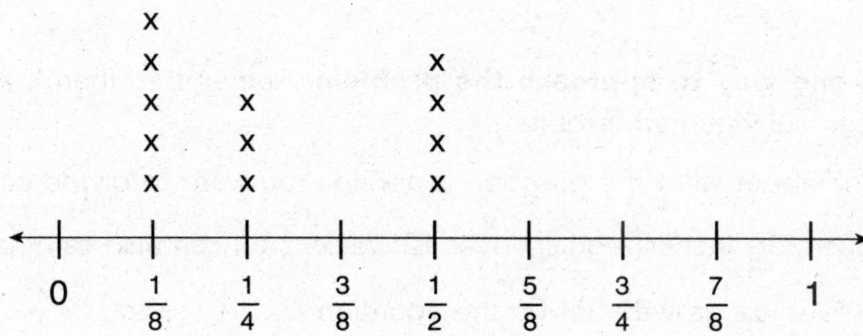

Weekly Growth of Bean Plant (in inches)

STEP 4: Now that you have recorded your data you can **analyze** the data.

HELPFUL HINT

Analyzing the data means that you can examine it to look for relationships, patterns, trends, etc.

What does the data say?

- Five out of the 12 weeks the plant grew $\frac{1}{8}$ in.
- Four out of the 12 weeks the plant grew $\frac{1}{2}$ in.
- Three out of the 12 weeks the plant grew $\frac{1}{4}$ in.
- The plant grew a total of $3\frac{3}{8}$ inches over 12 weeks.

$$\left[\left(5\times\frac{1}{8}\right)+\left(3\times\frac{1}{4}\right)+\left(4\times\frac{1}{2}\right)\right]=\frac{5}{8}+\frac{3}{4}+\frac{4}{2}=\frac{5}{8}+\frac{6}{8}+2=\frac{11}{8}+2=1\frac{3}{8}+2=3\frac{3}{8}$$

Here's something to think about.

You collected data for 12 weeks. You have identified that the plant grew $3\frac{3}{8}$ inches over the 12 weeks. Some weeks it grew more than other weeks. However, if the growth of $3\frac{3}{8}$ inches was distributed equally over the 12 weeks; how much did the plant grow every week?

Let's look at one way to approach the problem. Remember, there's always more than one way to solve a math problem.

STEP 1: Think about what the question is asking. You want to divide or share the growth, $3\frac{3}{8}$ inches, equally over 12 weeks. You can also call that finding the average growth. Set up the equation.

$$3\frac{3}{8}\div 12=\underline{\qquad\qquad}$$

You can call on your prior knowledge, number sense, and understanding of fraction division to help you solve this problem.

STEP 2: Remember mixed number fractions can be rewritten as

$$3\frac{3}{8}=3+\frac{3}{8}$$

and you also know that $\left(3+\frac{3}{8}\right)\div 12=3\frac{3}{8}\div 12$.

STEP 3: Let's solve the problem $\left(3+\frac{3}{8}\right)\div 12=(3\div 12)+\left(\frac{3}{8}\div 12\right)=$

Call on your knowledge again; you are dividing fractions, so solve each division problem.

$$3\div 12=3\times\frac{1}{12}=\frac{3}{12}=\frac{1}{4} \quad \text{and} \quad \frac{3}{8}\div 12=\frac{3}{8}\times\frac{1}{12}=\frac{3}{96}=\frac{1}{32}$$

Now let's add $\frac{1}{4}+\frac{1}{32}=\frac{8}{32}+\frac{1}{32}=\frac{9}{32}$. On average, the plant grew $\frac{9}{32}$ of an inch a week for 12 weeks.

Let's look at the second way to approach the problem.

STEP 1: Think about what the question is asking. You want to divide or share the growth, $3\frac{3}{8}$ inches, equally over 12 weeks. You can also call that finding the average growth. Set up the equation.

$$3\frac{3}{8}\div 12=\_\_\_\_\_\_\_\_\_\_$$

This time use your prior knowledge, number sense, and understanding of fractions to create an improper fraction.

STEP 2: Rewrite the problem by creating an improper fraction.

$$3\frac{3}{8}\div 12=\frac{27}{8}\div 12=\_\_\_\_\_\_\_\_\_\_$$

STEP 3: Divide (Use your knowledge of dividing fractions.)

$$\frac{27}{8}\div 12=\frac{27}{8}\times\frac{1}{12}=\frac{27}{96}=\frac{9}{32}$$

On average, the plant grew $\frac{9}{32}$ of an inch a week over 12 weeks. You get the same result using a different mathematical process.

I think you are ready to do a little experimenting of your own.

PRACTICE—CHECK FOR UNDERSTANDING: Interpreting Data

1. Use the dot plot below to answer the following questions.

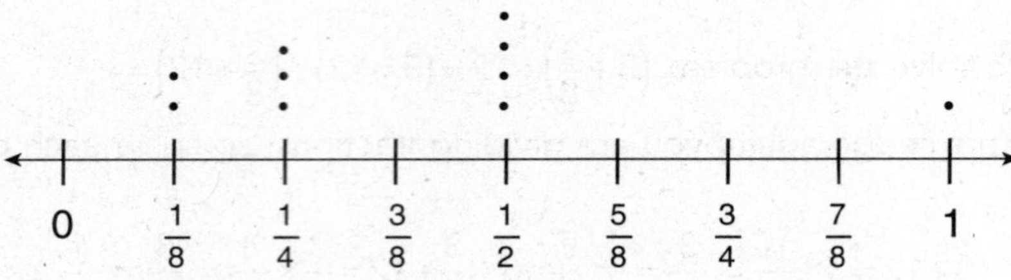

Snowfall Amounts (in inches) for Ten Days

A. What was the total snowfall for the 10 days? Write your answer in the box below.

B. What was the most frequent amount of snowfall in the 10 days?

____

What was the least frequent?

____

C. What is the difference, in actual snowfall amounts, between the most frequent snowfall and the least frequent snowfall?

2. Ms. Fisher's science class is studying the product packaging of potato chips compared to the labeling. The labeling on the bag states that it contains 32 oz of potato chips. Ms. Fisher and the class weighed 16 bags of potato chips to find that the content of each bag was slightly below the stated 32 oz weight.

Part A: Use the data below to create a line plot.

$\frac{1}{8}$	$\frac{1}{8}$	$\frac{1}{2}$	$\frac{1}{8}$
$\frac{1}{8}$	$\frac{1}{4}$	$\frac{1}{8}$	$\frac{1}{8}$
$\frac{1}{8}$	$\frac{1}{8}$	$\frac{1}{4}$	$\frac{1}{8}$
$\frac{1}{4}$	$\frac{1}{4}$	$\frac{1}{2}$	$\frac{1}{8}$

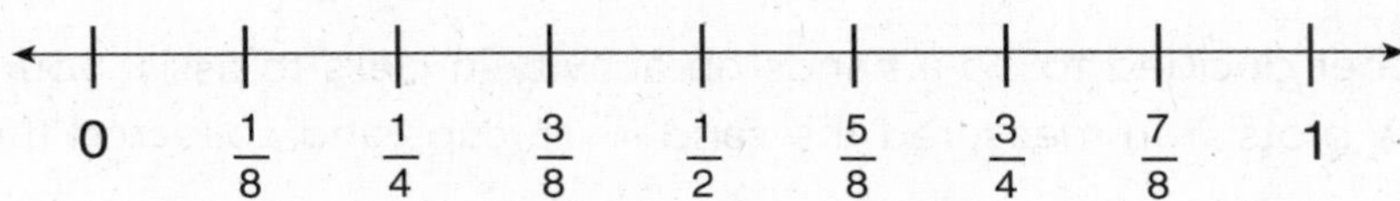

Difference in Potato Chip Bag Weight (oz)

Part B:

2.1 Which best describes how the bag weights are spread out? Circle the best description.

Clustered between 0 oz and $\frac{1}{4}$ oz	Clustered between $\frac{1}{4}$ oz and $\frac{1}{2}$ oz	Clustered between $\frac{1}{2}$ oz and $\frac{3}{4}$ oz

2.2 What is the average difference in weight for the 16 bags of chips?

[ ]

2.3 Identify whether the statements below are true or false.

Data Analysis		
There are $\frac{1}{5}$ as many $\frac{1}{2}$ oz bags as $\frac{1}{4}$ oz bags.	T	F
There are 5 times as many $\frac{1}{8}$ oz bags as $\frac{1}{2}$ oz bags.	T	F
Exactly $\frac{1}{8}$ of the total bags have a $\frac{1}{2}$ oz difference in weight.	T	F

2.4 Select the **two** fractions from the fractions below that represent the number of bags having less than a $\frac{1}{2}$ oz difference in weight when compared to the bag label.

☐ A. $\frac{5}{8}$

☐ B. $\frac{3}{4}$

☐ C. $\frac{7}{8}$

☐ D. $\frac{2}{3}$

☐ E. $\frac{14}{16}$

3. Your teacher decided to do a hands-on activity in class to assist you in learning about line plots. You measured the sand in 16 cups and collected the following data.

Sand Measurements (ft)

$\frac{1}{4}$	$\frac{1}{2}$	$\frac{1}{2}$	$\frac{1}{8}$
$\frac{1}{2}$	$\frac{1}{4}$	$\frac{1}{8}$	$\frac{3}{4}$
$\frac{1}{4}$	$\frac{3}{4}$	$\frac{1}{4}$	$\frac{1}{8}$
$\frac{1}{4}$	$\frac{3}{4}$	$\frac{1}{8}$	$\frac{1}{8}$

Part A: Use a number line to create a line plot to organize and display the data in the space below.

Part B: If you were to redistribute the sand so that each cup had the same amount, how much sand would be in each cup? Enter your answer and show your work in the space below.

Part C: Approximately how many **inches** of sand are in each cup? Select **one** choice below.

6 inches | 3 inches | 5 inches | 4 inches

Part D: Identify whether the statements below are true or false.

Data Analysis		
Five-eighths of the cups are filled with less than $\frac{1}{2}$ ft of sand.	T	F
There are 2 times as many cups filled with $\frac{1}{8}$ ft of sand as $\frac{3}{4}$ ft of sand.	T	F
The data is clustered between $\frac{3}{4}$ ft and 1 ft.	T	F

Answers (pages 206–209)

Question	Answer	Detailed Explanation
1	A. 4 inches B. Most $\frac{1}{2}$ in Least 1 in C. 1 in	A. $\frac{2}{8}+\frac{3}{4}+3$ C. $\left(4\times\frac{1}{2}\right)-1=2-1=1$
2A		X X X X X X X X X X above $\frac{1}{8}$; X X X X above $\frac{1}{4}$; X X above $\frac{1}{2}$ 0, $\frac{1}{8}$, $\frac{1}{4}$, $\frac{3}{8}$, $\frac{1}{2}$, $\frac{5}{8}$, $\frac{3}{4}$, $\frac{7}{8}$, 1 Weekly Growth of Bean Plant (in inches)
2B. 2.1	Clustered between 0 oz and $\frac{1}{4}$ oz	
2B. 2.2	$\frac{13}{64}$	$\frac{10}{8}+\frac{4}{4}+\frac{2}{2}=3\frac{2}{8}=3\frac{1}{4}$ $3\frac{1}{4}\div 16=3\frac{1}{4}\times\frac{1}{16}$
2B. 2.3	F	
	T	There are 2 bags that are $\frac{1}{2}$ oz below the stated 32 oz weight. $2\times 5=10$
	T	Two bags of the 16 bags are $\frac{1}{2}$ oz below the stated 32 oz weight. $\frac{2}{16}=\frac{1}{8}$
2B. 2.4	C, E	$\frac{14}{16}=\frac{7}{8}$ bags are less than $\frac{1}{2}$ oz difference in weight

Question	Answer	Detailed Explanation
3A		Line plot: x x x x x above $\frac{1}{8}$; x x x x x above $\frac{1}{4}$; x x x above $\frac{1}{2}$; x x x above $\frac{3}{4}$ 0, $\frac{1}{8}$, $\frac{1}{4}$, $\frac{3}{8}$, $\frac{1}{2}$, $\frac{5}{8}$, $\frac{3}{4}$, $\frac{7}{8}$, 1 Sand Measurements (ft)
3B	$\frac{45}{128}$ ft	$\left(\frac{5}{8}+\frac{5}{4}+\frac{3}{2}+\frac{9}{4}\right)\div 16=$ $\left(\frac{5}{8}+\frac{10}{8}+\frac{12}{8}+\frac{18}{8}\right)\div 16=$ $\frac{45}{8}\times\frac{1}{16}=\frac{45}{128}$
3C	4 inches	Several math applications come into play here, but you can handle it. 1. Remember how to convert between units of measurement. You have $\frac{45}{128}$ ft, how many inches do you have? You are converting from a larger unit to a smaller unit so you need to multiply by the conversion factor (1 ft = 12 inches). $\frac{45}{128}\times 12=4.218$ 2. You are multiplying a number by a fraction less than 1. What happens to your product? 3. You need to round the decimal to the nearest whole number.
3D		Five-eighths of the cups are filled with less than $\frac{1}{2}$ ft of sand. — T There are 2 times as many cups filled with $\frac{1}{8}$ ft as $\frac{3}{4}$ ft of sand. — F The data is clustered between $\frac{3}{4}$ ft and 1 ft. — F

Volume—Rectangular Prisms

THE STANDARDS

CCSS.5.MD.3. Recognize volume as an attribute of solid figures and understand concepts of volume measurement.

A. A cube with side length 1 unit, called a "unit cube," is said to have "one cubic unit" of volume and can be used to measure volume.

B. A solid figure that can be packed without gaps or overlaps using n unit cubes is said to have a volume of n cubic units.

CCSS.5.MD.4. Measure volumes by counting unit cubes, using cubic cm, cubic in, cubic ft, and improvised units.

CCSS.5.MD.5. Relate volume to the operations of multiplication and addition and solve real-world and mathematical problems involving volume.

A. Find the volume of a right rectangular prism with whole-number side lengths by packing it with unit cubes, and show that the volume is the same as would be found by multiplying the edge lengths, equivalently by multiplying the height by the area of the base. Represent threefold whole-number products as volumes (e.g., to represent the associative property of multiplication).

B. Apply the formulas $V = l \times w \times h$ and $V = b \times h$ for rectangular prisms to find volumes of right rectangular prisms with whole-number edge lengths in the context of solving real-world and mathematical problems.

C. Recognize volume as additive. Find volumes of solid figures composed of two nonoverlapping right rectangular prisms by adding the volumes of the nonoverlapping parts, applying this technique to solve real-world problems.

What Does This Mean?

Volume is a measure of the number of unit cubes it takes to fill a solid figure. To make sure you have a thorough understanding of volume, let's just talk about rectangular prisms for now. Imagine having a transparent cardboard box like the one in Figure 1 below. Yes, it's a square and you will learn why it's also a rectangle in Chapter 7.

The box has **dimensions**. It is long, it is wide, and it is tall or has height. As such, it is measured in cubic units (unit^3). The dimensions, length × width × height, create a space inside the box that can be measured by packing it with unit cubes. This stacking of unit cubes measures what is known as the volume.

Figure 1

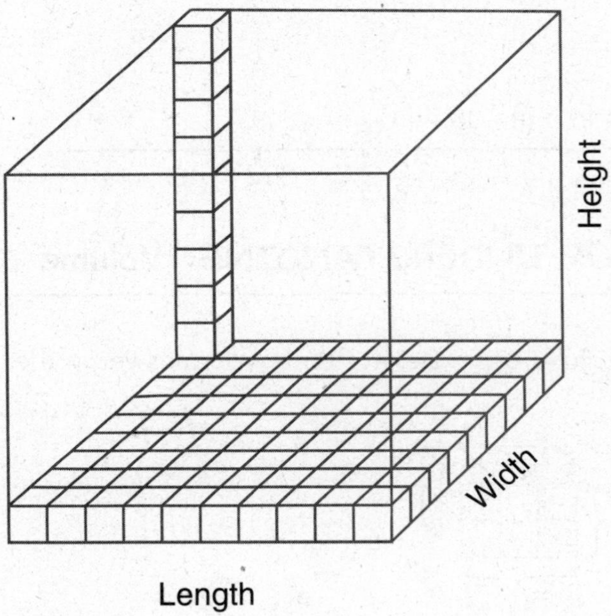

Height

Width

Length

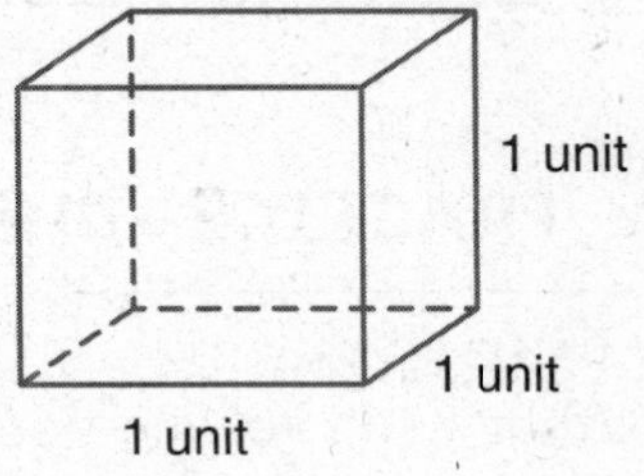

1 unit

1 unit

1 unit

A unit cube

HELPFUL HINT

A cube is a rectangular prism with sides of equal length.

Side A = Side B = Side C

Figure 2

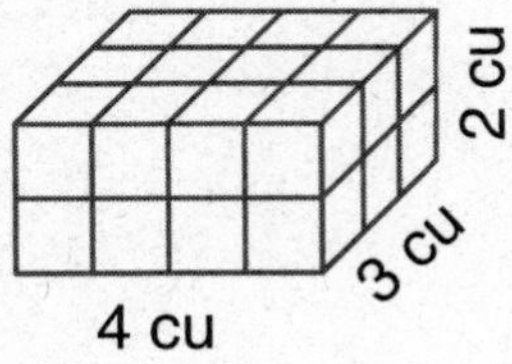

2 cu

3 cu

4 cu

Volume = length × width × height

$V = l \times w \times h$ or $V = b \times h$

$V = 4 \text{ cu} \times 3 \text{ cu} \times 2 \text{ cu} = 24 \text{ cu}^3$

Hopefully, you are having an Aha! moment, right? You just realized that you can find the volume of the prism, a solid rectangular figure, by counting the cubes. Figure 2 shows a rectangular prism with a base 12 square units (4 cu × 3 cu) and a height of 2 cu. See, this math thing is not as difficult as it looks, for now!

There are other solid figures, so at some point more volume formulas will come into play. However, right now focus on volume as being the space inside a solid figure which is measured in cubic units (cubic ft or ft^3, cubic inches or in^3, cubic cm or cm^3, etc.).

HELPFUL HINT

Think back to exponents, an in × in × in = in^3.

PRACTICE—CHECK FOR UNDERSTANDING: Volume

1. What is the volume of the rectangular prism below? Write your answer in the box below.

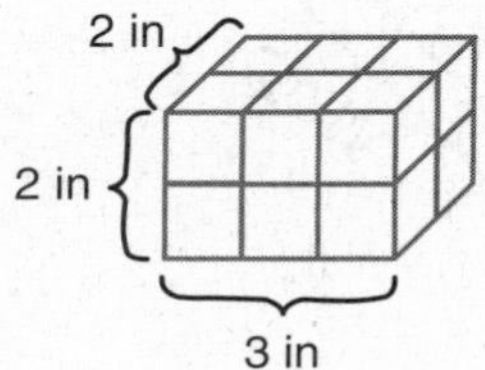

_____ cubic inches

2. What is the volume of the rectangular prism below?

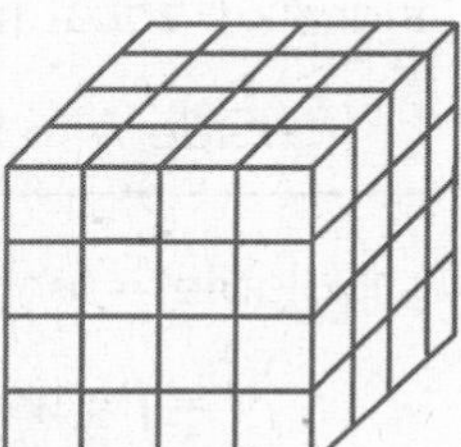

- A. 16 cubic units
- B. 36 cubic units
- C. 46 cubic units
- D. 64 cubic units

3. Find the volume of the rectangular prisms below in cubic units unless otherwise stated.

A. 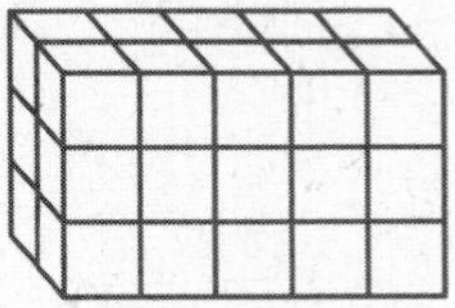	Answer:
B. 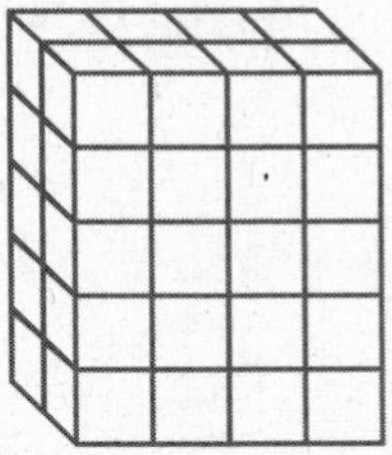	Answer:
C. 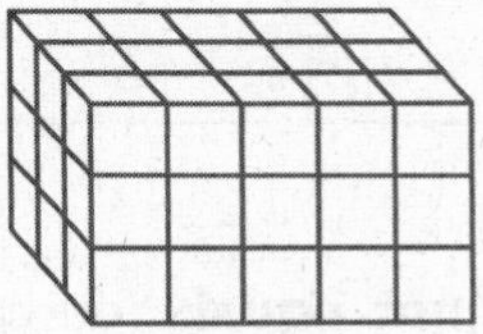	Answer:
D. 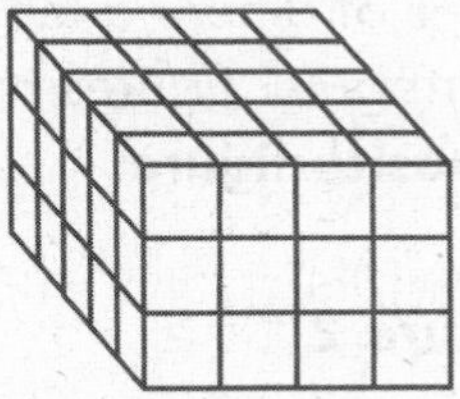	Answer:
E. 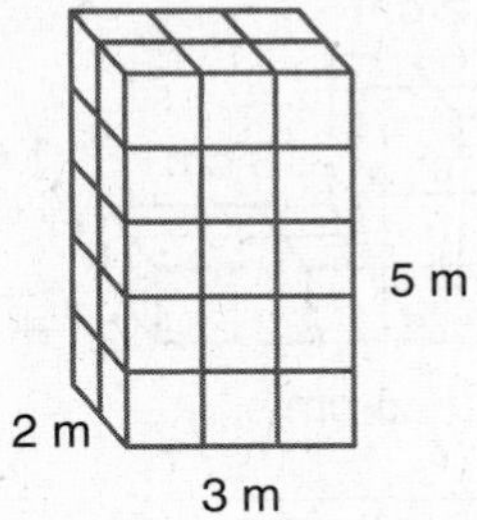 5 m, 2 m, 3 m	Answer:
F. 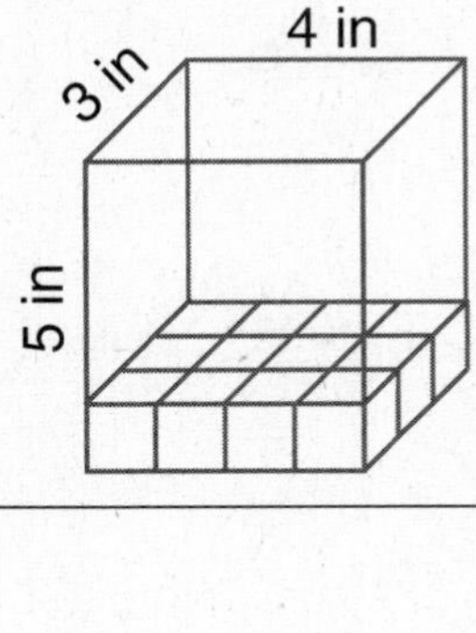 4 in, 3 in, 5 in	Answer:

Answers (pages 214–215)

Question	Answer
1	12 cubic inches
2	D
3	A. 30 cubic units
	B. 40 cubic units
	C. 45 cubic units
	D. 60 cubic units
	E. 30 m^3 or cubic meters
	F. 60 in^3 or cubic inches

Volume—Composite Figures

What Is an Irregular Figure?

A figure is considered to be irregular if the **sides are not equal**. You can see that Figure 1 has sides of equal lengths and Figure 2 has sides of differing lengths. Therefore, Figure 2 is considered to be an **irregular figure**. Irregular figures may be composed of two or more rectangular prisms called a **composite figure**.

Figure 1

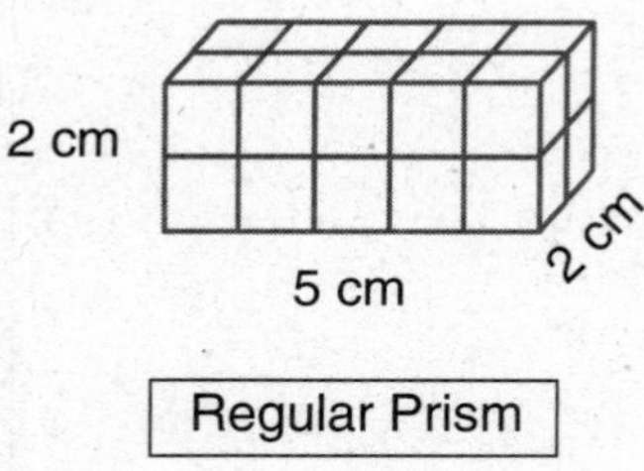

Figure 2

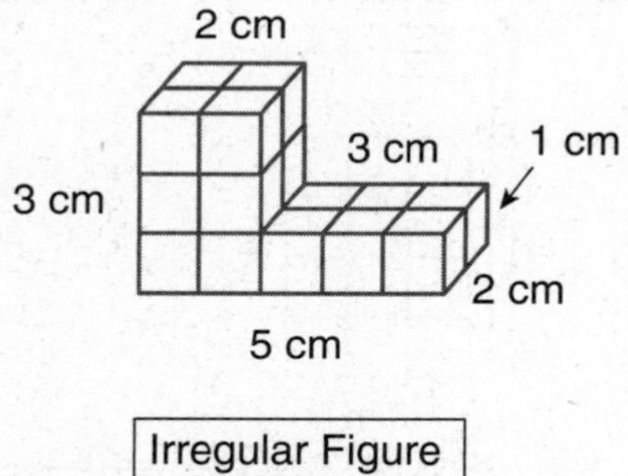

Taking a closer look at Figure 2, you can see that it is composed of two regular prisms. If you decompose the figures, you will have a regular prism with the dimensions 5 cm × 2 cm × 1 cm and another regular prism with the dimensions of 2 cm × 2 cm × 2 cm.

Regular Prism 1

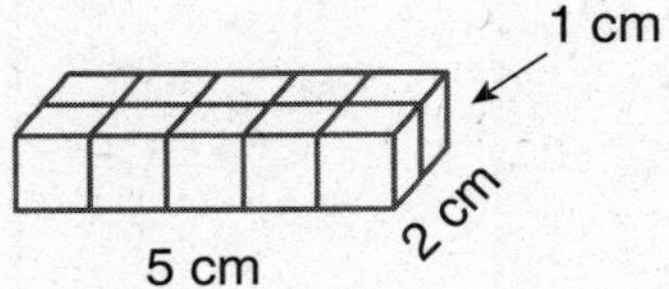

Regular Prism 2

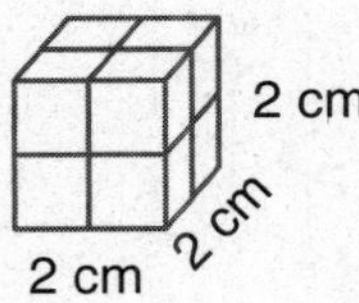

This should be another Aha! moment. Yes, to find the volume of an irregular figure you find the volume of each prism and then add the volumes together.

Volume Prism 1 = 5 cm × 2 cm × 1 cm = 10 cm^3

Volume Prism 2 = 2 cm × 2 cm × 2 cm = 8 cm^3

10 cm^3 + 8 cm^3 = 18 cm^3

Figure 2 has a volume of 18 cubic centimeters. To check your work just count the cubes. There are 18 unit cubes in Figure 2.

PRACTICE—CHECK FOR UNDERSTANDING: Volume—Irregular Figures

Find the volume of the irregular figures below in cubic units.

1. 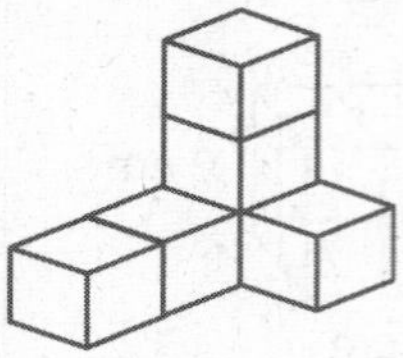	Answer:
2. 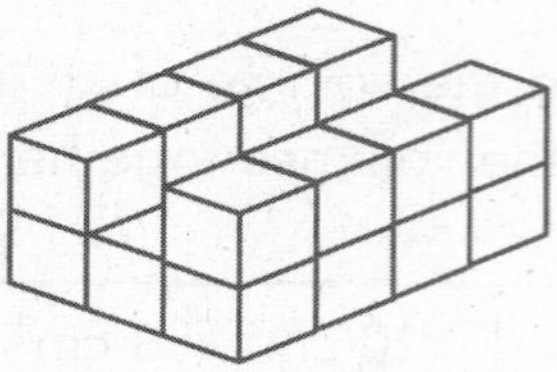	Answer:
3. 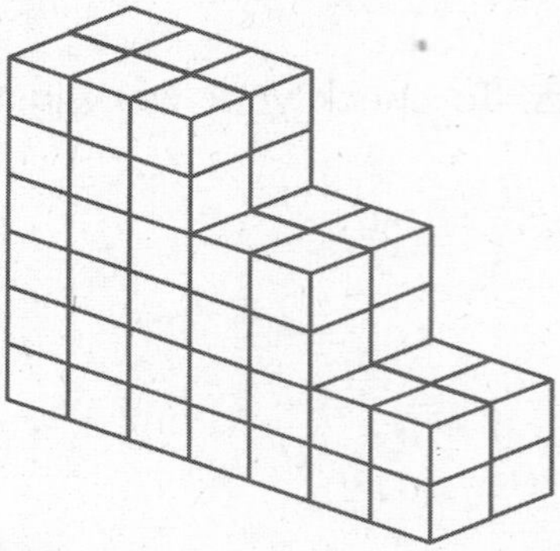	Answer:
4. 	Answer:

Answers (page 218)

Question	Answer
1	6 cubic units
2	20 cubic units
3	60 cubic units
4	140 cubic units

Volume—Problem Solving

1. Ramesh is a local architect that has been hired by the Townhouse Resort. He is designing the pool below for the resort. The diagram below shows the measurements of the pool.

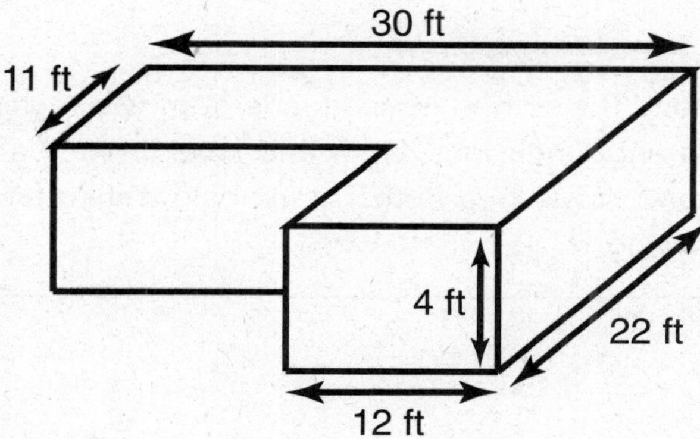

How many cubic feet of water are needed to fill the pool? Show your work in the space below:

2. A gardener is making two flower beds for her garden. Each flower bed will hold the same volume of dirt.

Part A: Two rectangular prisms form the first flower bed, shown below.

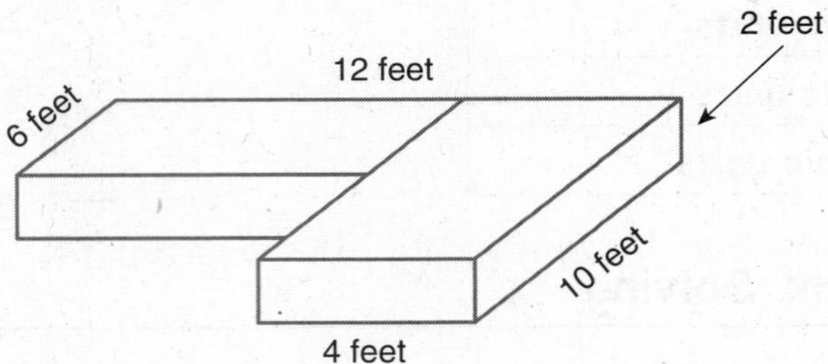

If she fills the bed up to the top with soil, how many cubic feet of soil will she need?

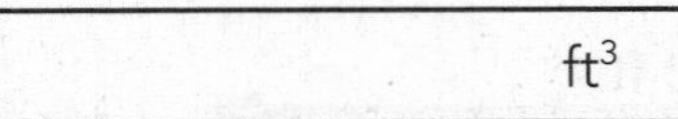

Part B: The second flower bed is also made from two rectangular prisms but has different dimensions. Draw and label (in feet) a possible design for the second flower bed in the space below, then find the volume of your drawing.

V =

3. Determine the amount of concrete needed to build the steps pictured below.

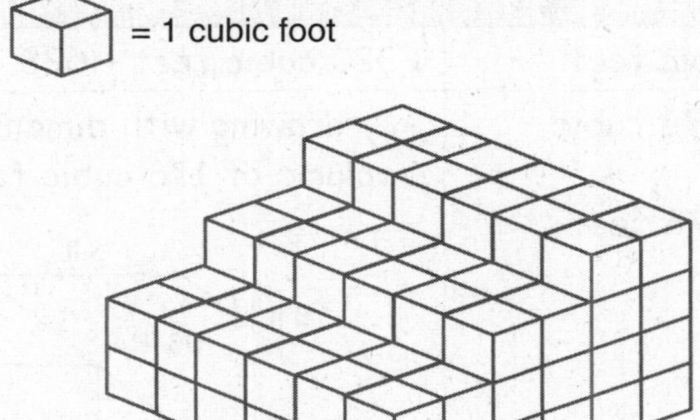

cubic feet

Answers (pages 219–221)

Question	Answer	Detailed Explanation
1	1,848 cubic feet	1,320 cubic feet + 528 cubic feet
2	Part A: 176 cubic feet Part B: 176 feet	Any drawing with dimensions that provide a volume of 176 cubic feet. For example, 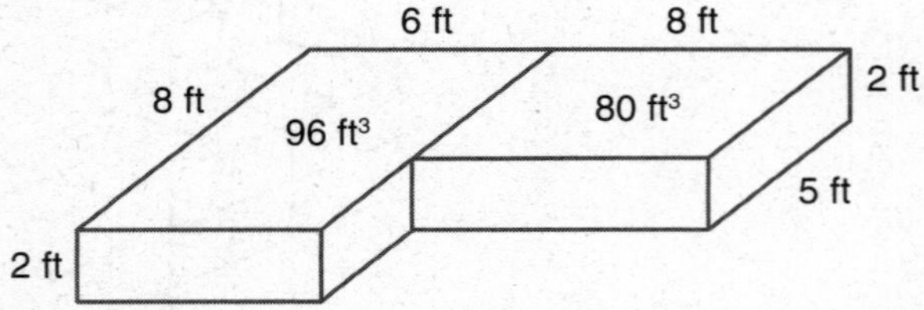 6 ft 8 ft 2 ft 8 ft 96 ft³ 80 ft³ 5 ft 2 ft
3	108 cubic feet	

Geometry

Have you heard the joke about the acorn? What did the acorn say when he was all grown up? Gee, om-a-tree!

Well rest assured that being a tree has very little to do with geometry. However, determining a position in space has a lot to do with geometry. Okay, what am I talking about? Ever played the game Battleship where you strategically place these plastic ships on a board that supposedly represents the sea? You call out shots like D-4, B-2 to identify if you have found the location of your opponent's ships in an effort to sink them? The objective of the game is that you have to sink your enemy's 5 ships in order to win. So why are you calling out D-4, B-2? Think about it, each call represents some point in space or, in this case, some point on the sea. Imagine this, if you begin to connect those points, you start to see things like lines and shapes.

My point, no pun intended! In this chapter, get ready to be a geometer, a mathematician who studies geometry. This chapter is sure to take you to different dimensions.

The Coordinate Plane: Graphing Points and Coordinate Values

THE STANDARDS

CCSS 5.G.1. Use a pair of perpendicular number lines, called axes, to define a coordinate system, with the intersection of the lines (the origin) arranged to coincide with the 0 on each line and a given point in the plane located by using an ordered pair of numbers, called its coordinates. Understand that the first number indicates how far to travel from the origin in the direction of one axis, and the second number indicates how far to travel in the direction of the second axis, with the convention that the names of the two axes and the coordinates correspond (e.g., *x*-axis and *x*-coordinate, *y*-axis and *y*-coordinate).

CCSS 5.G.2. Represent real-world and mathematical problems by graphing points in the first quadrant of the coordinate plane, and interpret coordinate values of points in the context of the situation.

What Does This Mean?

Well, remember you did some problem solving and graphing in Chapter 4. Now you are going to learn more about that game board I mentioned at the beginning of the chapter and what it actually represents, the **coordinate plane**. What is the coordinate plane? In thinking about the game, it's the playing board. In thinking about geometry, the coordinate plane is the space created by a pair of intersecting perpendicular number lines.

Let's look at some basic facts before you dive in. The first fact, a number line is straight and extends in both directions with no end. The number line has an **origin** labeled as zero, and each point on the number line represents its distance away from the origin. Another way to look at this, imagine your street being a number line, call your house the origin. The house to the right of you is the distance of one house from the origin, the house next to that house represents the distance of two houses from the origin. This particular number line is horizontal and known as the **x-axis**. As we move across the number line, each number on this number line can be referred to as an **x-coordinate**. Since the position of the point on this line is denoted by a single number, the line is considered one dimensional.

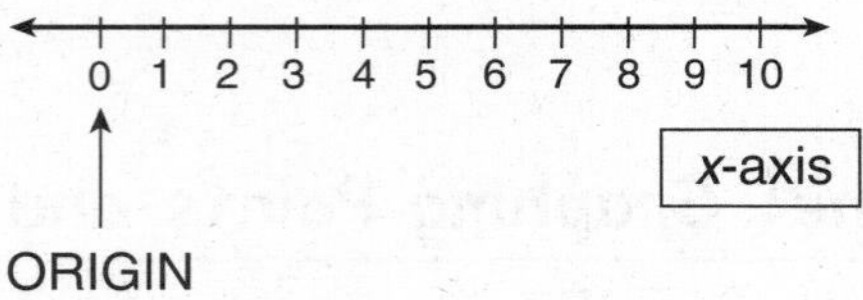

The second number line is vertical (perpendicular to the *x*-axis) and is known as the ***y*-axis**. This *y*-axis represents the street at the corner of your block. It intersects the *x*-axis at the origin zero. As we move up the line from the origin each number on this number line can be referred to as a ***y*-coordinate**.

It should be noted that each point in the coordinate plane on Graph 1 can be described by two numbers, the ***x*-coordinate** and the ***y*-coordinate**. These two numbers are also known as the ***x, y*-coordinate pair**. This coordinate pair describes the location of a point. On Graph 1, if you travel to the right of the origin 5 units across the *x*-axis and travel up three units on the *y*-axis you find yourself at the point (5, 3). Since the points on the coordinate plane are described by two numbers, the coordinate plane is considered to be **two-dimensional**. What is meant by two-dimensional? Imagine a piece of notebook paper, it is two-dimensional. Moving across the bottom of the paper represents your *x*-axis. Moving from the bottom of the paper to the top of the paper represents your *y*-axis.

Graph 1

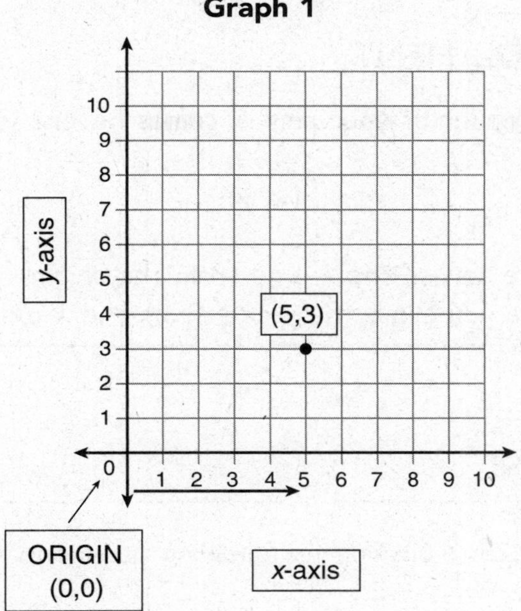

Looks very much like your game board, doesn't it?

Now that you understand what is meant by the coordinate plane and locating points, you can get to solving some real-world math problems. However, first let's just do a little practice.

Graph 2

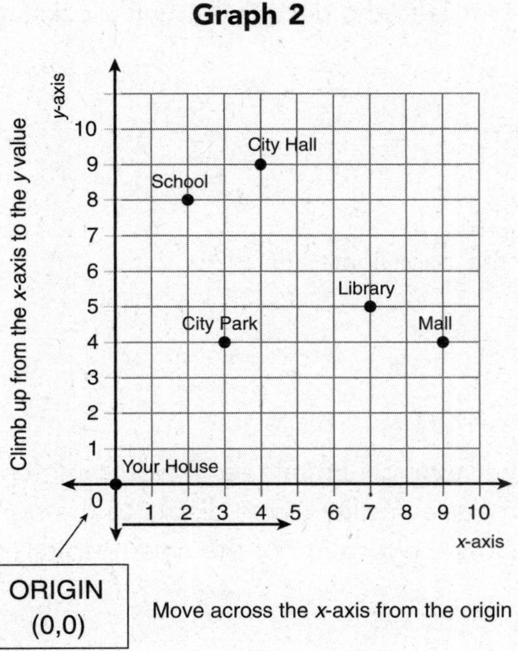

HELPFUL HINT

Remember the order of your pair, x comes before y.

$$(x, y)$$

x = the distance across the x-axis from the origin.
y = the distance you climb the y-axis from the x-axis.

YOUR TURN!

Use Graph 2 on page 225 to answer the following questions. *Note*: Each square represents a unit block.

1. The library is located at the coordinates (_____, ______).
- ○ A. (4, 3)
- ○ B. (7, 5)
- ○ C. (3, 4)
- ○ D. (2, 3)

2. You and friends decided to give the gaming station a rest and take a walk. If you walked 3 unit blocks to the right and 4 unit blocks up, where did you end up?
- ○ A. At the mall
- ○ B. At the library
- ○ C. At school
- ○ D. At the park

3. What building has the coordinates (4, 9)?
- ○ A. City Hall
- ○ B. The Park
- ○ C. The Mall
- ○ D. None of the above

4. You are on the City Planning Committee. The Committee has just approved the building of a new medical facility 5 unit blocks to the right of the park and 3 unit blocks up. What is the location of the new hospital?

(_____, ______)

5. Members of the community wanted the hospital to be built in a location closer to the public facilities (park, city hall, etc.). Name three possible locations where the medical facility could have been built.

A. (_____, _____)

B. (_____, _____)

C. (_____, _____)

6. Plot the points in question 5 on the coordinate plane below.

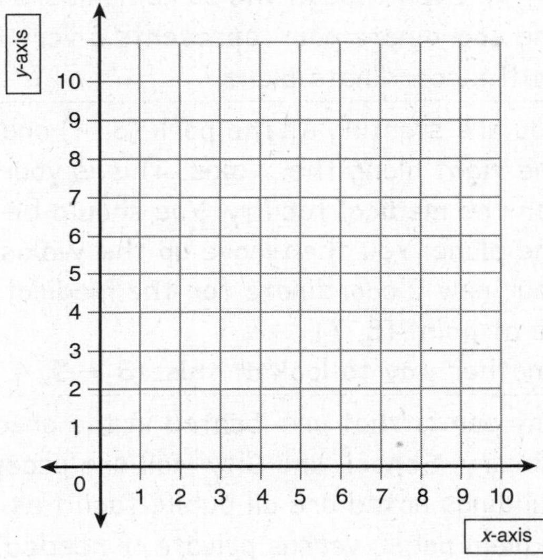

Answers (pages 226–227)

#	Answer	Detailed Explanation
1	B	(7, 5) is the coordinate pair. You move across the *x*-axis to 7 units and climb the *y*-axis 5 units where you reach the point that represents the location of the library.
2	D	You landed on the point representing the park, (3, 4)
3	A	Sometimes students confuse the order of the coordinate pair. Be careful, the mall represents the ordered pair (9, 4). Even though the same numbers are used, order of the coordinate pair represents a very different location on the coordinate plane.
4	(8, 7)	You are starting at the park (3, 4) and moving 5 units to the right along the *x*-axis. This is your new *x*-coordinate for the medical facility. You should be at point (**8**, 4) on the plane. You then move up the *y*-axis 3 units. This is your new *y*-coordinate for the medical facility. You should be at point (8, **7**). Another way to look at this. (3 + **5**, 4 + **3**) = (8, 7)
5	Answers may differ.	Any points that are located in the area of the Park, Library, School, and City Hall are acceptable. The buildings noted are all public facilities. Teacher can explain public versus private if needed. Sample points: (4, 6), (5, 7)
6	Sample points (4, 6), (5, 7). Accept any sample points from question 5 plotted correctly on the plane.	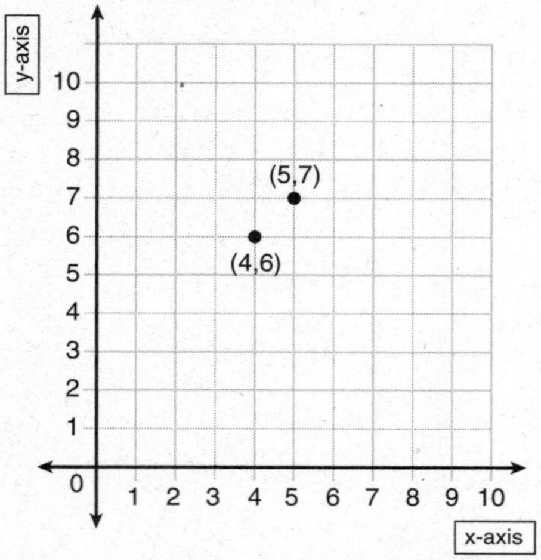 y-axis 10 9 8 7 6 5 4 3 2 1 0 1 2 3 4 5 6 7 8 9 10 x-axis (5,7) (4,6)

PRACTICE—CHECK UNDERSTANDING: The Coordinate Plane

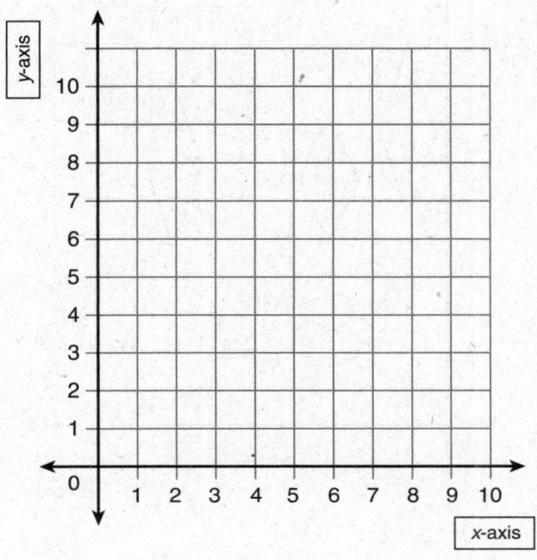

1. Plot and label the point in the table below on the coordinate plane above.

Point	x	y
Q	2	8
R	3	3
S	0	5
T	7	0
U	8	4
V	6	7

2. You are playing the board game Battleship with a friend. The coordinate plane below shows the location of your ships.

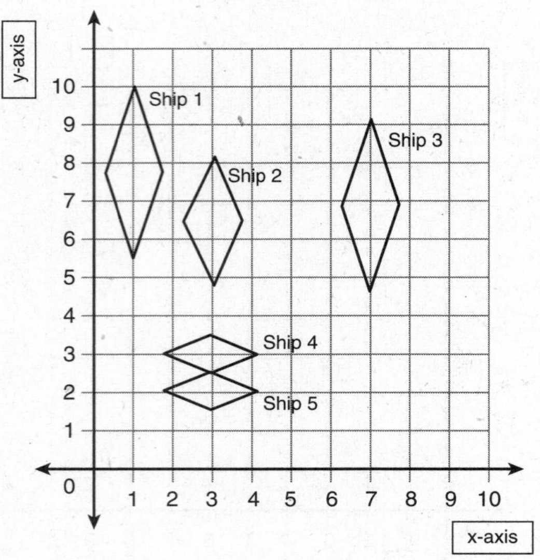

Determine if your opponent has hit or sunken any of your ships.

Part A. Plot the following points on the coordinate plane above:

Opponent Calls	*x*	*y*
1	2	4
2	3	3
3	1	7
4	7	6
5	8	4
6	7	7
7	4	1
8	1	9
9	3	2
10	3	9
11	1	8
12	2	2

Part B. The rules show that ships with three points are sunk, with two points are badly damaged, with one point or no points are still in battle. Circle the result for each ship.

Ship #	Results			
1	Sunk	Badly Damaged	In battle	No hits
2	Sunk	Badly Damaged	In battle	No hits
3	Sunk	Badly Damaged	In battle	No hits
4	Sunk	Badly Damaged	In battle	No hits
5	Sunk	Badly Damaged	In battle	No hits

Part C. How many complete misses were there? ________________

Name the coordinate pairs that were complete misses (there may be more boxes than coordinate pairs).

(___, ___)	(___, ___)	(___, ___)
(___, ___)	(___, ___)	(___, ___)

Answers (pages 229–231)

1.

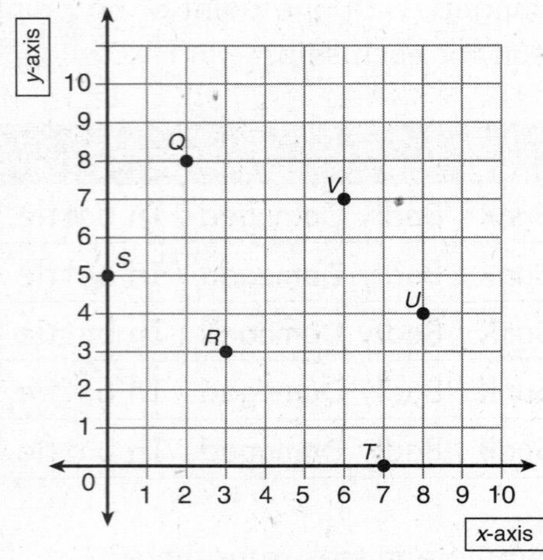

Point	x	y
Q	2	8
R	3	3
S	0	5
T	7	0
U	8	4
V	6	7

2. **Part A.**

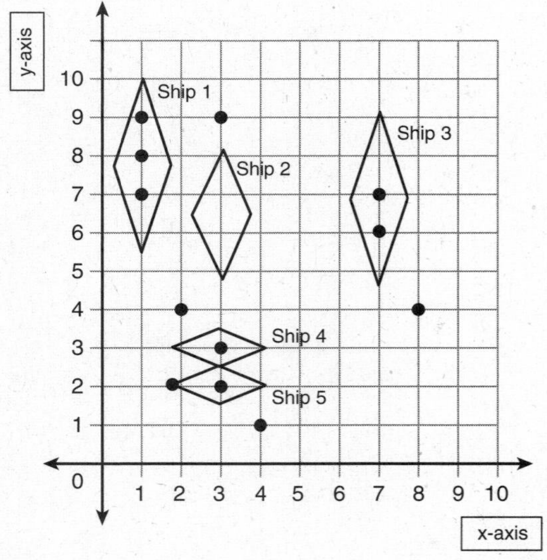

Part B.

Ship #	Results			
1	Sunk			
2			In battle	No hits
3		Badly Damaged		
4			In battle	
5		Badly Damaged		

Part C. There were four complete misses: (2, 4), (3, 9), (4, 1), (8, 4).

Classifying Two-Dimensional Figures

THE STANDARDS

CCSS 5.G.3. Understand that attributes belonging to a category of two-dimensional figures also belong to all subcategories of that category. For example, all rectangles have four right angles, and squares are rectangles, so all squares have four right angles.

CCSS 5.G.4. Classify two-dimensional figures in a hierarchy based on properties.

What Does This Mean?

When you look at a shape, let's say a rectangle, how do you know a rectangle is a rectangle? What makes a rectangle a rectangle? Is it the number of sides it has? Is it the number of angles it has? Is it the shape's angle measurements? Is it the shape's number of parallel lines?

Once you understand the properties of a rectangle you can then tackle the question, Is a square a rectangle? If you base your response on the appearance of each figure below placed side by side, I'm sure your answer is NO!!! However, are you sure?

What makes a rectangle a rectangle? Does a square share any properties or characteristics of a rectangle? Let's move forward and find out.

What Is a Triangle?

By definition you can say a triangle is a **polygon** (a closed figure) with three straight sides and three angles. However, not all triangles are the same. Triangles can be classified by the length of the sides or by the measurement of the angles.

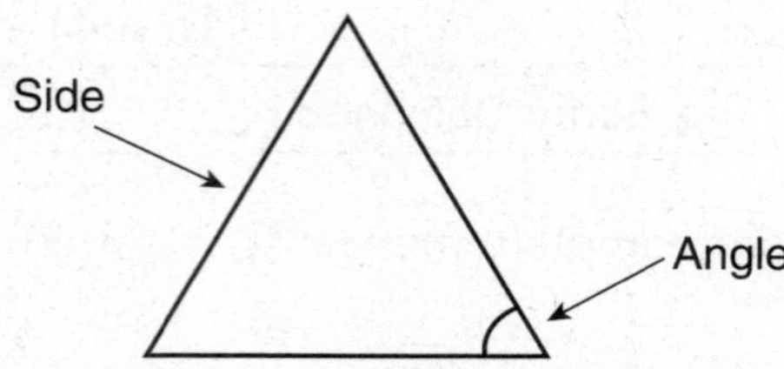

Naming Triangles Based Upon the Sides

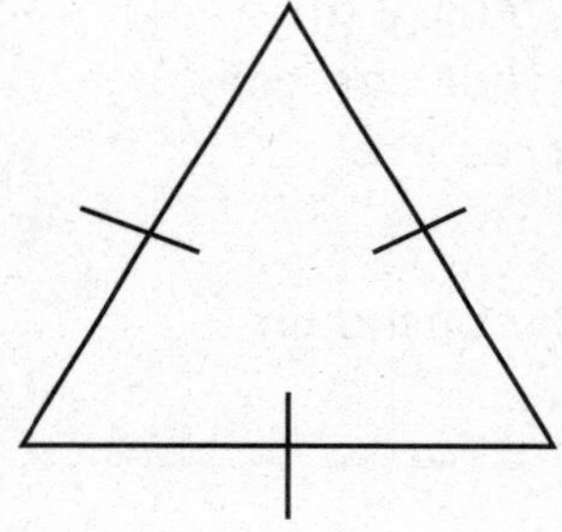

Key Concept

Equilateral Triangle—All three sides are the same length. The tick marks on the figure to the left identify that each side has an equal length of measurement. All three sides are *congruent*, meaning they are the exact same length. All three angles are *congruent*, having an equal measure of 60° (the symbol ° = degrees).

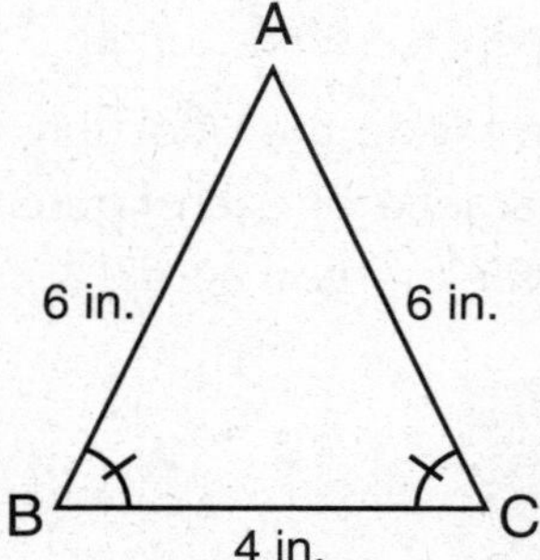

Key Concept

Isosceles Triangle—Two sides have equal lengths or are *congruent*. The angles opposite the equal sides are congruent. The figure to the left has two sides both measuring 6 inches. The tick marks on the angles opposite the equal sides identify the angles are *congruent*, having the same measure.

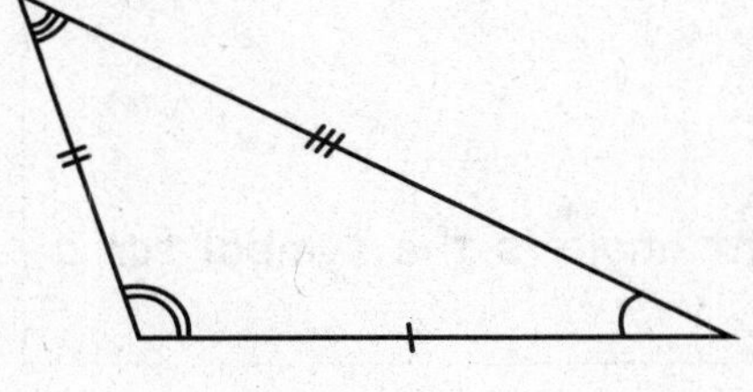

Key Concept

Scalene Triangle—Each side has a different length of measurement. The tick marks on the sides of figure to the left identify that each side has a different measurement. Each angle in a scalene triangle also has three different measures.

Naming Triangles Based upon the Angles

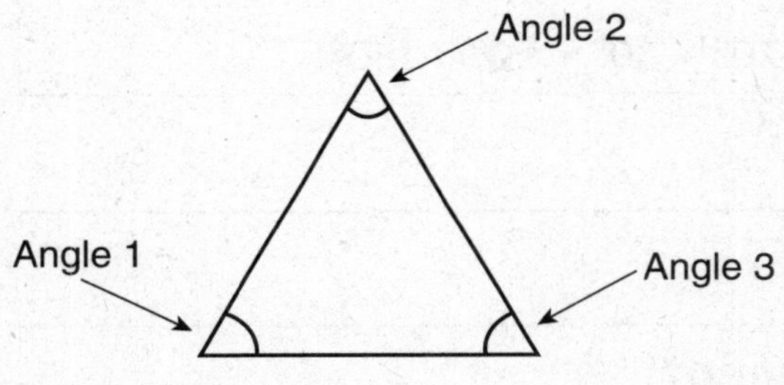

HELPFUL HINT

The angle measures of a triangle will always add up to 180° (the symbol ° = degrees).

Angle 1 + Angle 2 + Angle 3 = 180°

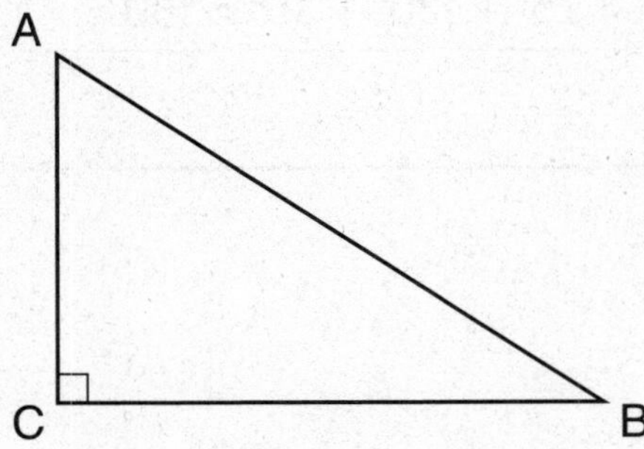

Key Concept

Right Triangle—This triangle has one angle measuring exactly 90°.

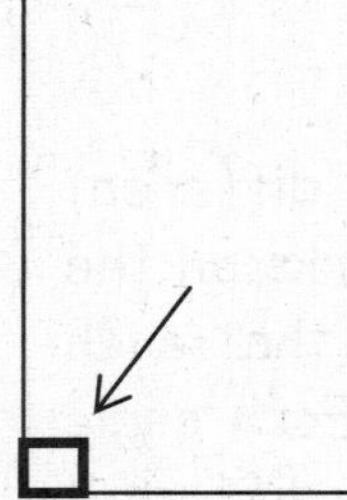

HELPFUL HINT

The square in the corner of a right angle is the symbol for a right angle = 90°.

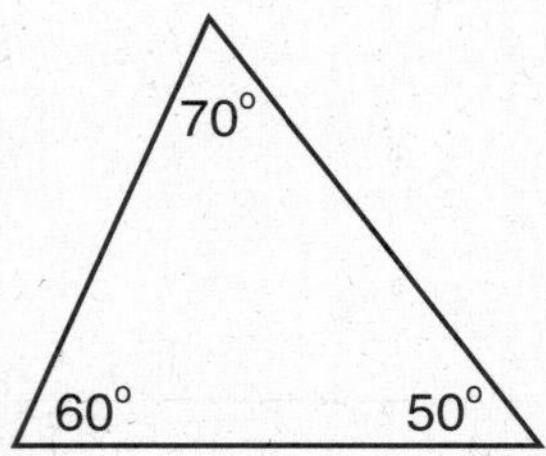

Key Concept

Acute Triangle—Any angle measuring less than 90° is considered to be an acute angle. An acute triangle has three angles measuring less than 90°.

Total Angles Sum = 70° + 60° + 50° = 180°

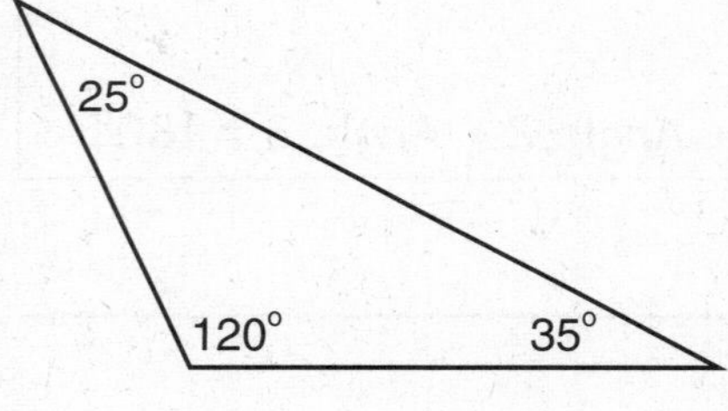

Key Concept

Obtuse Triangle—Any angle measuring more than 90° is considered to be an obtuse angle. An obtuse triangle has one angle that measures more than 90°.

Total Angles Sum = 25° + 120° + 35° = 180°

Combining the two methods of classifying triangles creates triangles with special characteristics.

	Acute	Right	Obtuse
Equilateral	60°, 60°, 60°		
Isosceles	38°, 71°, 71°	45°, 45°	40°, 100°, 40°
Scalene	46°, 71°, 63°	40°, 50°	26°, 23°, 131°

What Is a Quadrilateral?

By definition you can say a quadrilateral is a **polygon** (a closed figure) with four straight sides and four angles. However, not all quadrilaterals are the same either. Quadrilaterals are also classified by the length of the sides or by the measurement of the angles.

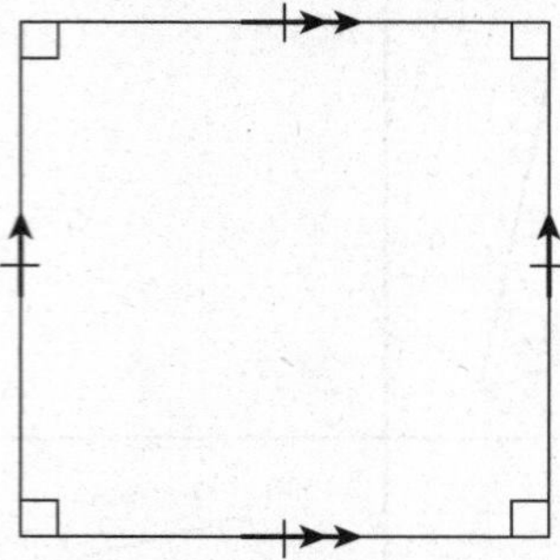

Key Concept

Square A square has four *congruent* (equal) sides and four *congruent* angles, all are right angles. Remember, a right angle measures 90° degrees.

HELPFUL HINTS

1. The angle measures of a quadrilateral will always add up to 360° (the symbol ° = degrees).

   Total angles sums 90° + 90° + 90° + 90° = 360°

2. The opposite sides of a square are also parallel; meaning the opposite sides are always an equal distance from each other and will never intersect or touch (→ or →→ indicates parallel sides).

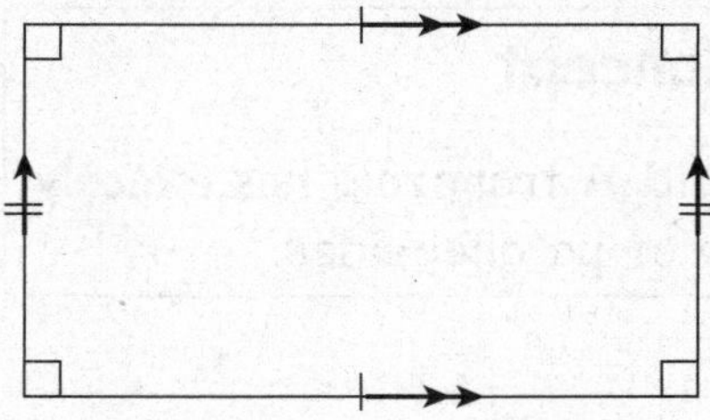

Key Concept

Rectangle—A rectangle has two pairs of **congruent (equal)** sides and four right angles.

Total angles sums 90° + 90° + 90° + 90° = 360°

HELPFUL HINT

The opposite sides of a rectangle are parallel, meaning the opposite sides are always an equal distance from each other and will never intersect or touch.

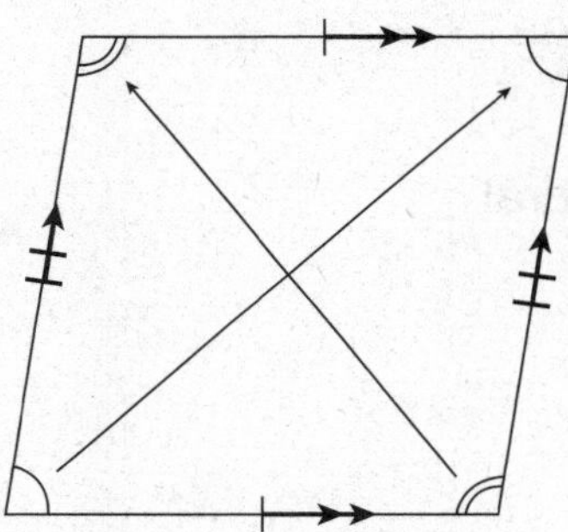

Key Concept

Parallelogram—A parallelogram has two pairs of congruent sides that are **parallel** and two pairs of congruent angles that are opposite each other.

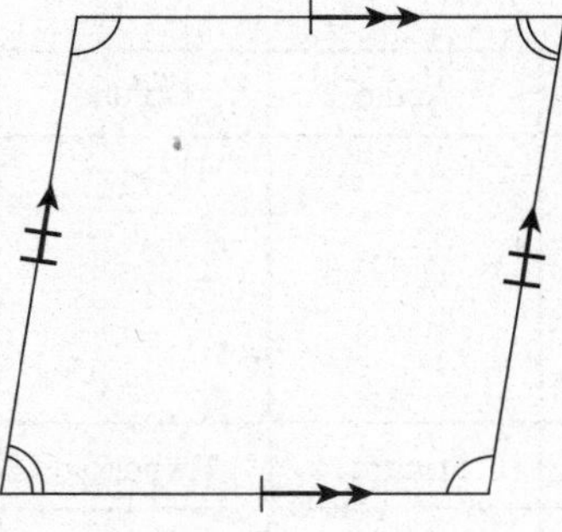

Key Concept

Rhombus—A rhombus has four equal sides. It also has two pairs of parallel sides and two pairs of congruent angles that are opposite each other.

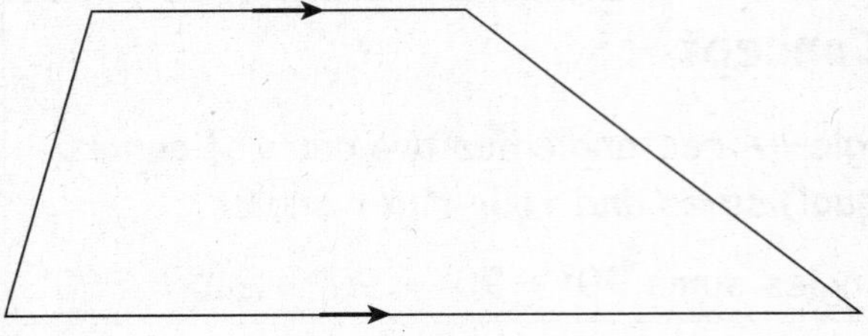

Key Concept

Trapezoid A trapezoid has exactly one pair of parallel sides.

Hierarchy of Triangles and Quadrilaterals

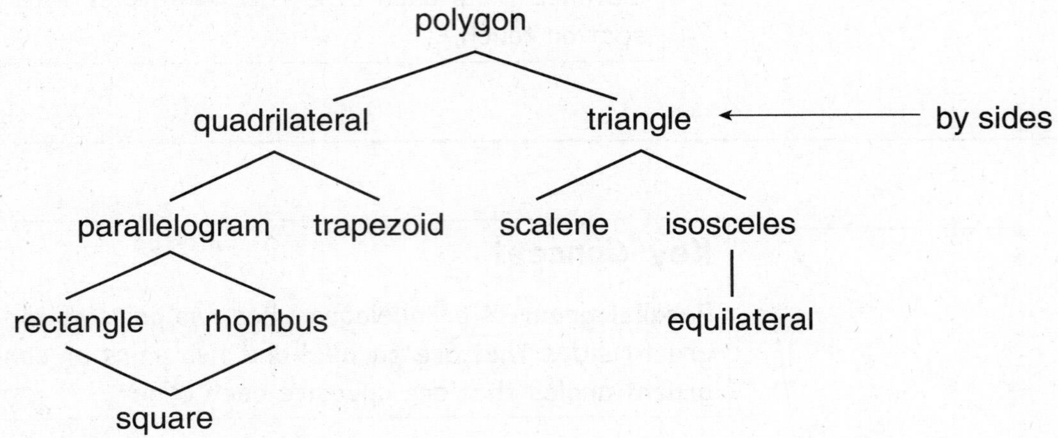

YOUR TURN!

Check all the properties that can describe each shape in the table.

Shape	Equilateral	Isosceles	Scalene	Right	Acute	Obtuse
1.						
	Quadrilateral	Parallelogram	Rectangle	Rhombus	Square	Trapezoid
2.						

Answers (page 240)

Question	Answer	Detailed Explanation
1	Isosceles, right	**Isosceles**: Two sides have equal lengths or are congruent. The angles opposite the equal sides are congruent. **Right:** Note the square in the corner means the angle measures 90°.
2	Quadrilateral, parallelogram, rectangle, rhombus, square	Review the hierarchy of quadrilaterals. A square meets all properties except that of a trapezoid. quadrilateral parallelogram trapezoid rectangle rhombus square

PRACTICE—CHECK UNDERSTANDING: Triangles and Quadrilaterals

1. Draw a line to match the shape to its classification.

Shape	Classification
	Rhombus
	Obtuse, Scalene
	Rectangle
	Isosceles, Acute

2. Use the following words to complete the chart:

Trapezoid	Rectangle	Square	Quadrilateral	Parallelogram	Rhombus

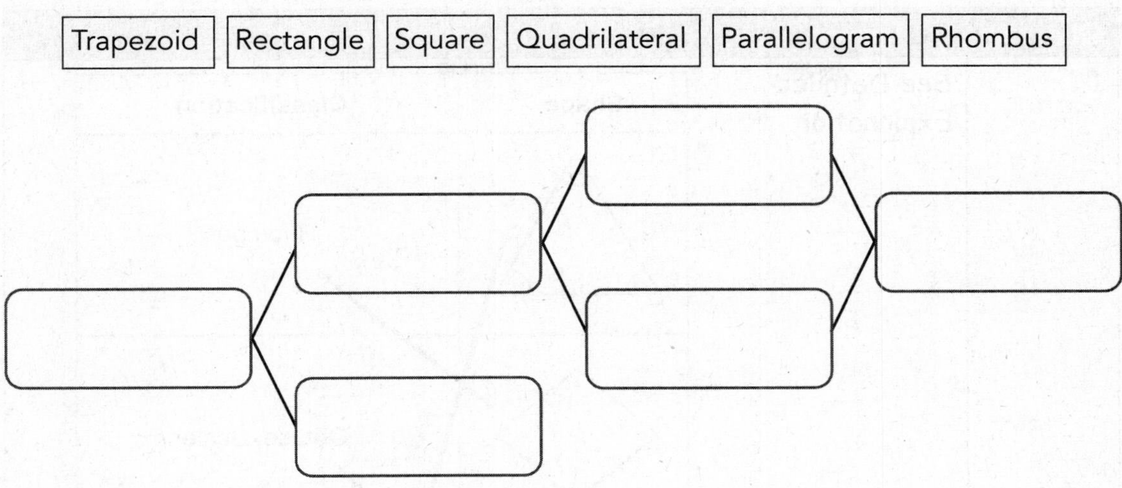

3. Select the **two** statements that are true about triangles.

- ☐ A. Isosceles triangles can be classified as equilateral triangles.
- ☐ B. Equilateral triangles can be classified as isosceles triangles
- ☐ C. Isosceles triangles have all the properties of equilateral triangles.
- ☐ D. Equilateral triangles have all the properties of isosceles triangles.

4. The cafeteria in your school is a quadrilateral with 4 right angles and 2 pairs of parallel sides. What type of quadrilateral is it? (Select all that apply)

- ☐ A. Rectangle
- ☐ B. Trapezoid
- ☐ C. Rhombus
- ☐ D. Square

Answers (pages 242–243)

Question	Answer	Detailed Explanation
1	See Detailed Explanation.	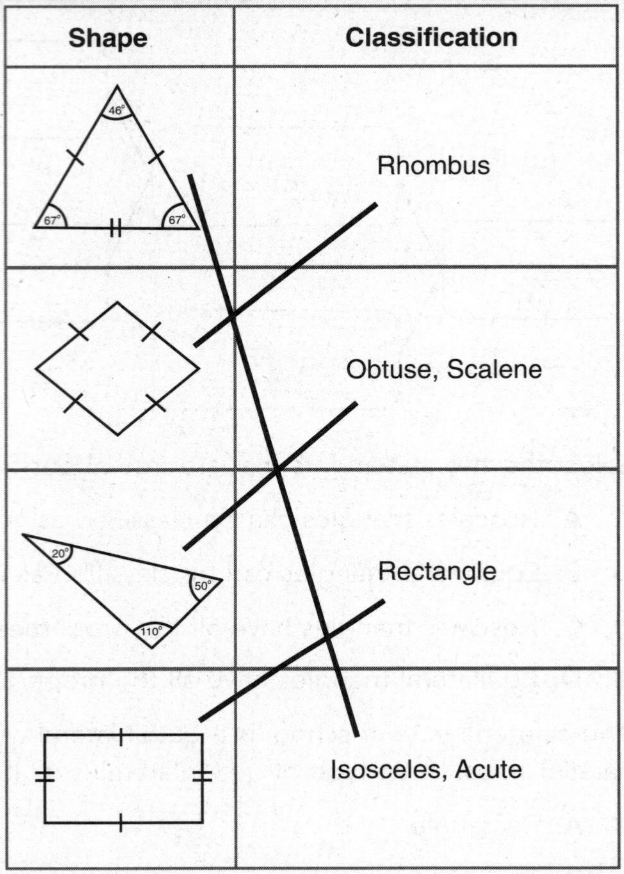 Shape / Classification: Rhombus; Obtuse, Scalene; Rectangle; Isosceles, Acute (triangle with angles 46°, 67°, 67° — Isosceles, Acute; rhombus — Rhombus; triangle with angles 20°, 50°, 110° — Obtuse, Scalene; rectangle — Rectangle)
2	See Detailed Explanation.	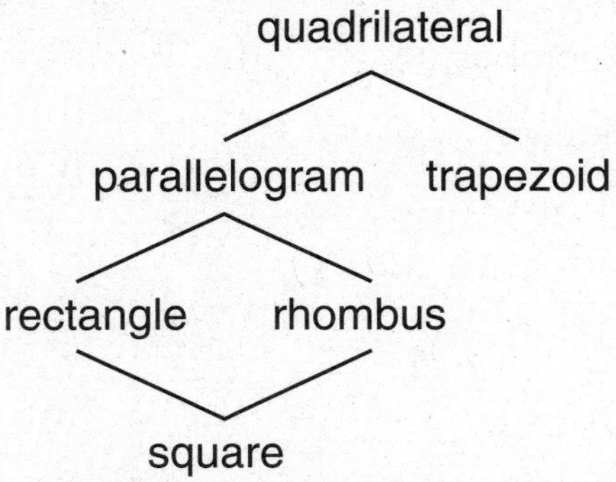 quadrilateral → parallelogram, trapezoid; parallelogram → rectangle, rhombus; rectangle, rhombus → square
3	B, D	An isosceles triangle has two pairs of equal sides. Since an equilateral triangle has three equal sides, it has at least two pairs of equal sides and can be classified as an isosceles triangle (see figure on page 242).
4	A, C, D	See detailed explanation above for answer #2.

CHAPTER 8
Practice Test 1

The following practice test contains Type 1, Type 2, and Type 3 questions similar to those that may appear on the PARCC assessment. **You may not use a calculator on this practice assessment.** The questions will be assigned various point values depending on their difficulty and how many separate components you will need to answer. Please refer to the information in Chapter 1 if you need more clarification on the PARCC or view The Mathematics High Level Blueprint at *http://www.parcconline.org/assessments/test-design/mathematics/math-test-specifications-documents* to identify the total number of tasks and/or items types and the point values for the PARCC.

IMPORTANT NOTE

Barron's has made every effort to create sample tests that accurately reflect the PARCC Assessment. However, the tests are constantly changing. The following two tests differ in length and content, but each will still provide a strong framework for fifth-grade students preparing for the assessment. Be sure to consult *http://www.parcconline.org* for all the latest testing information.

PRACTICE TEST 1

Read all problems carefully.

1. What is the volume of the rectangular prism in cubic units?

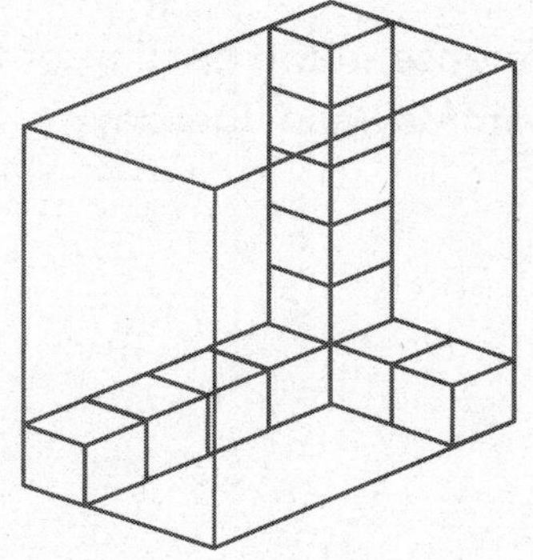

L × W × H
3 × 5 × 6

2. Which of these are equal to 78.684? Select the **two** correct answers.

- ☐ A. seventy-eight and six hundred eighty-four thousandths
- ☐ B. $(7 \times 10) + (8 \times 1) + (6 \times 1/10) + (4 \times 1/1{,}000)$
- ☐ C. Seven hundred eight and sixty-eight and four hundred thousandths
- ☐ D. $(7 \times 10) + (8 \times 1) + (6 \times 1/10) + (8 \times 1/100) + (4 \times 1/1{,}000)$

3. Select the **two** comparisons that are correct.

- ☐ A. $(5 \times 10) + (6 \times 1/100) < 50.11$
- ☐ B. Nineteen and seventy-eight hundreds = 19.078
- ☐ C. Four and twenty-eight thousandths > 4.023
- ☐ D. $21.14 < (2 \times 10) + (1 \times 1) + (3 \times 1/100)$

4. Malia plays for the local basketball team. She scored $\frac{1}{8}$ of her team's points on Tuesday and $\frac{1}{3}$ of her team's points on Friday. Her team scored the same number of points in both games.

Part A: What fraction of team's points did she score in both games?

- ○ A. $\frac{3}{8}$
- ○ B. $\frac{13}{24}$
- ○ C. $\frac{5}{12}$
- ○ D. $\frac{11}{24}$

Part B: She also played a game on Saturday where she scored $\frac{5}{9}$ of the points. What is the difference in the fraction of the total number of points she scored on Saturday and Tuesday?

- ○ A. $\frac{3}{7}$
- ○ B. $\frac{1}{2}$
- ○ C. $\frac{31}{72}$
- ○ D. $\frac{40}{72}$

5. Josh cooked gumbo. 15 pots of gumbo required 8 lb of seafood. He also made 10 deli sandwiches using 7 lb of shrimp.

Part A: Each sandwich was made with the same amount of shrimp. Which fraction represents the closest amount of shrimp in each sandwich?

○ A. $\frac{1}{2}$ lb

○ B. $\frac{3}{4}$ lb

○ C. $\frac{5}{6}$ lb

○ D. $\frac{2}{9}$ lb

Part B: Each pot of gumbo has the same amount of seafood. About how much seafood is in each pot?

○ A. $\frac{3}{4}$ lb

○ B. $\frac{7}{10}$ lb

○ C. $\frac{1}{2}$ lb

○ D. $\frac{3}{12}$ lb

6. Here is an expression:

$$\frac{4}{6} + \frac{1}{3}$$

Which of the following expressions below have the correct common dominators to add the two fractions? Select the **two** correct answers.

□ A. $\frac{8}{12} + \frac{2}{12}$

□ B. $\frac{8}{18} + \frac{12}{18}$

□ C. $\frac{12}{18} + \frac{6}{18}$

□ D. $\frac{4}{6} + \frac{2}{6}$

□ E. $\frac{7}{9} + \frac{6}{9}$

7. A toy manufacturer produces 480 special edition toys every month. The special edition set has three sizes:

   - The Junior set contains 50 pieces, and $\frac{5}{8}$ of the shipment will contain Junior sets.
   - The Advanced set contains 80 pieces, and $\frac{1}{3}$ of the shipment will be Advanced sets.
   - The Masters' Edition contains 200 pieces, and the rest of the shipment will be Masters' Editions.

   Part A: Determine how many of each set are to be shipped a month.

   Part B: Determine how many pieces, in total, need to be created.

Enter your answers, your work, and/or explanation in the space provided.

8. Mary is reporting for the school newspaper. She stated that two track team members ran less than a $\frac{1}{2}$ mile in 2 minutes. Runner A ran 880 yards and Runner B ran 860 yards (1 mile = 1,760 yards).

Part A: Compare a $\frac{1}{2}$ mile to the distance of each jogger using the >, <, or = symbol. Explain whether Mary's reasoning is correct for Runner A and Runner B. Enter your comparisons and your explanation in the space provided.

Part B: She also reported the team's long jump distances shown below.

Distance Jumper A: $15\frac{3}{4}$ ft   Distance Jumper B: $18\frac{1}{4}$ ft

She reported that Jumper B jumped $3\frac{2}{4}$ ft farther than Jumper A because the difference between the whole numbers is 3 and the difference between the numerators is 2.

- Explain why Mary's reasoning is incorrect.
- What is the correct difference, in feet, between the two jumpers distances?

Enter your explanation and your answer in the box below.

9. Ahmed is trying to determine the distance he will drive over the next two days. He drove $\frac{3}{4}$ of a mile today and plans on driving $\frac{5}{8}$ of a mile tomorrow. He wrote:

$$\frac{3}{4}+\frac{5}{8}=\frac{8}{12}$$

- Explain why Ahmed's answer is not reasonable.
- Find the total distance, in miles, Ahmed will drive over the two days.
- Explain how to use the number line to show your answer is correct.

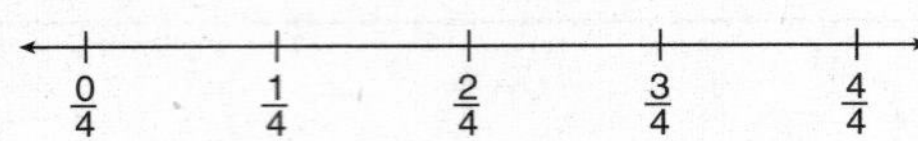

Enter your work and your explanations in the space provided.

10. Which statement can be represented by the expression $\frac{5}{6}\times\frac{3}{8}$?

○ A. $\frac{5}{6}\times\frac{3}{8}$ is 6 groups of $\frac{3}{8}$, divided into 5 equal parts.

○ B. $\frac{5}{6}\times\frac{3}{8}$ is 5 groups of $\frac{3}{8}$, divided into 6 equal parts.

○ C. $\frac{5}{6}\times\frac{3}{8}$ is 5 groups of $\frac{3}{8}$, divided into 48 equal parts.

○ D. $\frac{5}{6}\times\frac{3}{8}$ is 6 groups of $\frac{3}{8}$, divided into 15 equal parts.

11. Which expression is equal to $\frac{2}{9}$?

○ A. $9 \div 2$

○ B. 9×2

○ C. $2 \div 9$

○ D. $9 - 2$

12. Answer Part A and Part B.

Part A: Select **two** equations that are correct when the number 30 is entered in the blank.

□ A. $300 \div$ ___ $= 15$

□ B. ___ $\times 12 = 300$

□ C. ___ $\div 3 = 10$

□ D. $2{,}400 \div$ ___ $= 80$

Part B: Select **two** equations that are correct when the number 400 is entered in the blank.

□ A. $80{,}000 \div$ ___ $= 200$

□ B. $30 \times$ ___ $= 1{,}200$

□ C. ___ $\times 400 = 160{,}000$

□ D. $1700 \times$ ___ $= 2{,}100$

13. The fifth-grade class at Martin Elementary School has 384 students. The entire class voted on 5 activities for the end-of-year school trip. Surprisingly, only $\frac{1}{4}$ of the class voted for Activity 1—Amusement Park, $\frac{3}{8}$ of the class voted for Activity 2—Broadway Play, $\frac{1}{6}$ of the class voted for Activity 3—Beach, and of the remaining students, half voted for Activity 4—Movies and the other half for Activity 5—Professional Baseball Game.

Part A: Determine how many students voted for each activity. Complete the table below:

Activity	Number of Students
Amusement Park	
Broadway Play	
Beach	
Movies	
Pro Baseball Game	

Part B: Broadway play tickets cost $25 a student. The theater is running a special, $10 off the regular ticket price for students younger than 10 years old.

- $\frac{3}{8}$ of the students going to the play are younger than 10.
- What is the total price the school pays for tickets to the Broadway play?
- Explain how you determined the number of students younger than 10 and the total price of the tickets.

14. Each small cube in this irregular figure measures 1 unit on each side.

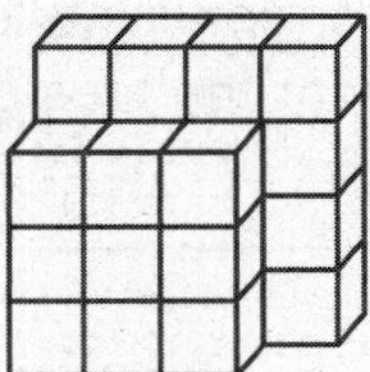

- What is the volume of the figure?
- Explain how you found the volume. You may show your work in your explanation.
- If 7 additional unit cubes were added to the original figure, what would be the new dimensions?
- Explain how you determined the dimensions of the new figure.

15. John is building a rectangular patio that is 15 feet long and 17 feet wide.

Part A: Write an equation that represents the area of John's patio. In your equation, let *y* represent the area of the patio. Then solve your equation.

Part B: John will be putting a fence around three sides of the patio to block his neighbors' view of his yard. One foot of fence costs $57. Write an expression to represent the total cost of the fence. Explain how you determined your expression. Enter your expression and explanation in the space provided.

Part C: Use your expression from Part B to find the total cost, in dollars, of the fence. Enter your answer in the space provided.

16. The teacher drew an area model to find the value of 1,703 ÷ 9.

Teacher's Model for 1,703 ÷ 9

	R	80	T	W
9	900	S	81	

- Determine the number that each letter in the model represents and explain each of your answers.
- Write the quotient and remainder for 1,703 ÷ 9.
- Explain how to use multiplication to check that your quotient is correct. You may show your work in your explanation.

Enter your answers and explanations in the space provided.

PRACTICE TEST 1—ANSWERS EXPLAINED

Question	Answer	CCSS	Detailed Explanation
1	90 cubic units	5.MD.3	
2	A, D	5.NBT.3a	
3	A, C	5.NBT.3b	
4	Part A: D Part B: C	5.NF.2	
5	Part A: B Part B: C	5.NF.3	
6	C, D	5.NF.1	
7	A. 300 Junior sets, 160 Advanced sets, 20 Masters' Editions B. 31, 800 pieces	5.NF.4 and 5.NF.6	A. Junior: $480 \times \frac{5}{8} = 300$ Advanced: $480 \times \frac{1}{3} = 160$ 300 + 160 = 460 Masters': 480 – 460 = 20 B. $300 \times 50 = 15,000$ $160 \times 80 = 12,800$ $20 \times 200 = 4,000$ 31,800
8	Part A: $\frac{1}{2}$ of 1,760 = 880 Part B: $2\frac{1}{2}$ ft	4.NF.2	880 = 880 860 < 880 Mary's reasoning is incorrect about Runner A. Runner A ran the equivalent of a $\frac{1}{2}$ mile. Mary was supposed to subtract the $15\frac{3}{4}$ from $18\frac{1}{4}$. She could have changed the fractions to improper fractions $\frac{73}{4} - \frac{63}{4}$. She also could have skip counted up by fourths on the number line until she reached $18\frac{1}{4}$.

Question	Answer	CCSS	Detailed Explanation
9	The distance he will drive in 2 days is $1\frac{3}{8}$ miles.		Ahmed's answer is not reasonable. $\frac{3}{4}$ is more than $\frac{1}{2}$ and $\frac{5}{8}$ is slightly more than $\frac{1}{2}$. The two added together will be more than 1. Ahmed's answer is less than one whole. You can use benchmark fractions on the number line to show that each number is more than $\frac{1}{2}$ and approximate that the two numbers added together are more than 1 on the number line. You can also show that $\frac{8}{12}$ is greater than $\frac{1}{2}$ but less than 1.
10	B	5.NF.4a	
11	C	5.NF.3	
12	Part A: C, D Part B: A, C	5.NBT. Int.1	
13	Part A: Amusement Park, 96 Broadway Play, 144 Beach, 64 Movies, 40 Pro Baseball Game, 40 Part B: $3,060	5.NF.6	$384 \times \frac{1}{4} = 96$ $384 \times \frac{3}{8} = 144$ $384 \times \frac{1}{6} = 64$ 304 384 – 304 = 80 The number of remaining students $\frac{1}{2} \times 80 = 40$ $144 \times \frac{3}{8} = 54$ under the age of 10 144 – 54 = 90 older than 10 90 × $25 = $2,250 54 × $15 = $810 Total paid = $3,060

Question	Answer	CCSS	Detailed Explanation
14	*V* = 25 cubic units New Dimension 4 unit × 2 unit × 4 unit	5.MD.5a	Possible answers to explain how: Counted squares or decomposed figures and found volume of each prism. Figure was already 25 cubic units. I just added the 7 unit cubes to 25 and got 32.
15	Part A: *y* = 15 × 17 *y* = 255 sq ft Part B: Fence 1: [2(15) + 17] × $57 Fence 2: [2(17) + 15] × $57 Part C: Fence 1 [2(15) + 17] × $57 = $2,679 Fence 2 [2(17) + 15] × $5 = $2,793	4.MD.3	17 15 15 17 15 17 17 15 There are two possible ways to fence the patio. Either is acceptable. Fence 1: 15 ft × 17 ft × 15 ft Fence 2: 17 ft × 15 ft × 17 ft

Question	Answer	CCSS	Detailed Explanation
16	R = 100 S = 720 T = 9 W = 2	5.NBT.6	9)1703 − 900 100 = R 803 S = − 720 80 83 − 81 9 = T 2 = W Quotient 189 r 2 $9 \times 100 = 900$ $9 \times 80 = 720$ $9 \times 9 = 81$ Total = 1,701 + 2 = 1,703

Practice Test 2

CHAPTER 9

PRACTICE TEST 2

The following practice test contains Type 1, Type 2, and Type 3 questions similar to those that may appear on the PARCC assessment. **You may not use a calculator on this practice assessment.** The questions will be assigned various point values depending on their difficulty and how many separate components you will need to answer. Please refer to the information in Chapter 1 if you need more clarification on the PARCC or view The Mathematics High Level Blueprint at *http://www.parcconline.org/assessments/test-design/mathematics/math-test-specifications-documents* to identify the total number of tasks and/or items types and the point values for the PARCC.

IMPORTANT NOTE

Barron's has made every effort to create sample tests that accurately reflect the PARCC Assessment. However, the tests are constantly changing. The following two tests differ in length and content, but each will still provide a strong framework for fifth-grade students preparing for the assessment. Be sure to consult *http://www.parcconline.org* for all the latest testing information.

Read all questions carefully.

1. The number 700,000 can also be expressed as

- ○ A. $7 \times 10 \times 10 \times 10 = 7 \times 10^3$
- ○ B. $7 \times 10 \times 10 \times 10 \times 10 \times 10 = 7 \times 10^5$
- ○ C. $7 \times 10 \times 10 \times 10 \times 10 = 7 \times 10^4$
- ○ D. $7 \times 10 \times 10 = 7 \times 10^2$

2. Select the **two** ways to represent the number

701.269

☐ A. Seven hundred and one hundred sixty-nine hundredths

☐ B. 700 + 1 + 0.2 + 0.06 + 0.009

☐ C. $(70 \times 100) + (1 \times 10) + \left(3 \times \frac{1}{10}\right) + \left(6 \times \frac{1}{100}\right) + \left(9 \times \frac{1}{100}\right)$

☐ D. Seven hundred one and two hundred sixty-nine thousandths

3. Place the appropriate comparison symbol <, =, > in the box.

$(5 \times 10) + (6 \times 1) + \left(8 \times \frac{1}{10}\right) + \left(3 \times \frac{1}{100}\right) \ \square \ (5 \times 10) + (6 \times 1) + \left(8 \times \frac{1}{10}\right) + \left(3 \times \frac{1}{100}\right) + \left(9 \times \frac{1}{1{,}000}\right)$

4. Enter your answer in the box.

$296 \times 35 =$ ☐

5. Enter your answer in the box.

$427 \times 2{,}087 =$ ☐

6. Enter your answer in the box.

$5.7 \times 0.1 =$ ☐

$5.7 \div 0.1 =$ ☐

7. Enter your answer in the box.

$9.68 + 245.37 =$ ☐

8. Enter your answer in the box.

$3.41 \times 1.6 =$ ☐

9. Enter your answer in the box.

$5.2 \div 0.4 =$ ☐

10. Enter your answer in the box.

$\frac{7}{12} - \frac{1}{4} =$ ☐

11. Enter your answer in the box.

$\frac{2}{3} + \frac{1}{4} + \frac{3}{8} =$ ☐

12. Enter your answer in the box.

$\frac{3}{5}+\frac{1}{2}-\frac{3}{4}=$ [ ]

13. Kaitlyn subtracted the fractions below:

$7\frac{5}{6}-2\frac{1}{3}=$ [ ]

Which **two** responses show how Kaitlyn solved the problem?

☐ A. Kaitlyn subtracted the whole numbers 7 – 2 and $\frac{5}{6}-\frac{2}{6}$ to arrive at an answer of $5\frac{1}{2}$.

☐ B. Kaitlyn subtracted the whole numbers 7 – 2 and $\frac{6}{9}-\frac{2}{9}$ to arrive at an answer of $5\frac{4}{9}$.

☐ C. Kaitlyn subtracted the whole numbers 7 – 2 and $\frac{15}{18}-\frac{6}{18}$ to arrive at an answer of $5\frac{9}{18}$.

☐ D. Kaitlyn subtracted the whole numbers 7 – 2 and $\frac{5}{6}-\frac{3}{6}$ to arrive at an answer of $5\frac{1}{3}$.

14. Your classmate arrived at the following solution for the problem below:

$$\frac{1}{3}+\frac{5}{6}=\frac{6}{9}$$

Which **one** of the following answers provides the best reason why the equation is incorrect?

○ A. $\frac{6}{9}>\frac{5}{6}$

○ B. $\frac{6}{9}<\frac{5}{6}$

○ C. $\frac{6}{9}=\frac{1}{3}$

○ D. $\frac{6}{9}<\frac{5}{2}$

15. Beverly needs to make $4\frac{3}{4}$ pounds of fudge for 2 math classes before Friday.

- On Monday she made $\frac{5}{8}$ pounds of fudge.
- On Tuesday she made $1\frac{1}{6}$ pounds of fudge.

Part A: What is the amount of fudge made on the two days? Enter your answer in the space provided.

Part B: Tuesday afternoon the principal asked Beverly to make an additional $1\frac{1}{3}$ pounds of fudge for another math class. How many remaining pounds of fudge does Beverly need to make before Friday? Enter your answer in the space provided.

16. Which **one** of the following answers provides the best explanation of the fraction below:

$$\frac{2}{5}$$

- ○ A. Two hot dogs divided equally among four friends.
- ○ B. Five hot dogs divided equally among two friends.
- ○ C. Two hot dogs eaten by one friend and not shared with friends.
- ○ D. Two hot dogs divided equally among five friends.

17. Fifteen students sold an equal amount of candy for the school candy sale. The order was delivered in 350 boxes.

Part A: How many boxes of candy should each person get? Enter your answer in the space provided.

Part B: Between what two whole numbers does your answer lie? Enter your answer in the space provided.

________ and ________

18. In gym, you were required to jump rope for $\frac{1}{4}$ of an hour every day, for five days. Select the equation that correctly identifies how many hours you jumped rope in 5 days.

- ○ A. $\frac{1}{4} \times 5 = 1 \times 5 \div 4 = \frac{5}{4}$ hours
- ○ B. $\frac{1}{4} \div 5 = 1 \div 4 \times 5 = \frac{1}{20}$ hours
- ○ C. $\frac{4}{1} \times 5 = 4 \times 5 = 20$ hours
- ○ D. $\frac{1}{4} + 5 = \frac{6}{4} = 1\frac{1}{2}$ hours

19. The fifth-grade class went to the circus. That day $\frac{7}{8}$ of the class had cotton candy. Later in the day $\frac{2}{3}$ of the cotton candy eaters felt sick. What fraction of the class fell ill? Write your answer in the box below.

20. Select a phrase to correctly fill the blank of each sentence.

GREATER THAN LESS THAN EQUAL TO

The product of $1\frac{1}{4}$ and 3 is ________________ 3.

The product of $\frac{2}{3}$ and 5 is ________________ 5.

The product of $\frac{2}{2}$ and $\frac{3}{4}$ is ________________ $\frac{3}{4}$.

21. Abigail read $\frac{1}{8}$ of a novel that was $\frac{2}{3}$ inch thick. How much of the book (in inches) did she read?

- O A. $\frac{1}{2}$
- O B. $\frac{3}{11}$
- O C. $\frac{1}{12}$
- O D. $\frac{4}{18}$

22. Satesh makes $9 an hour working at Game Stop. Pay day is next week. Due to the Thanksgiving weekend he worked $32\frac{2}{3}$ hours. How much will his check be next week? Write your answer in the box below.

23. Kathy uses felt fabric to make school pennants. Kathy has 21 yards of felt fabric. She uses $\frac{1}{3}$ yard of felt fabric to make each pennant. What is the total number of pennants Kathy can make with all 21 yards of fabric?

- O A. 63 pennants
- O B. 24 pennants
- O C. 7 pennants
- O D. 72 pennants

24. Complete the conversion by filling in the empty box with the correct number below:

0.05	50,000	0.5	5,000	50

50 cm = ☐ m

5 km = ☐ m

0.5 km = ☐ cm

25. The school cafeteria made 15 gallons of punch for the fifth grade holiday party. You and two friends had 32 ounces of punch with your lunch, when Ms. Lucille, the cafeteria aide, recognized that she had accidently served 5 quarts of punch at lunch, including your glass.

Part A: How much punch is left for the party? Write your answer in the box below.

☐

Part B: Each student is allowed to have 1.25 pints of punch at the party. How many students will the remaining punch serve? Write your answer in the box below.

☐

26. You are doing a science experiment and have collected the following data for the rainfall over the last 12 days.

$\frac{1}{2}$ inch	$\frac{1}{4}$ inch	$\frac{1}{4}$ inch
$\frac{1}{4}$ inch	$\frac{1}{4}$ inch	$\frac{1}{8}$ inch
$\frac{1}{4}$ inch	$\frac{1}{8}$ inch	$\frac{1}{8}$ inch
$\frac{3}{4}$ inch	$\frac{1}{2}$ inch	$\frac{1}{2}$ inch

Part A: Use the number line below to create a line plot to display the data.

0 1

Part B: What was the average rainfall over the 12 days period? Show your work in the box below.

27. The rectangular prism shown is made from cubes. Each cube is 1 cubic centimeter.

What is the volume of the rectangular prism shown? Write your answer in the box below.

cubic cm

28. A chest has a height of 38 inches and a base that is 27 inches long and 22 inches wide. What is the volume of the toy chest in cubic inches?

- ○ A. 15,756 cubic inches
- ○ B. 32,125 cubic inches
- ○ C. 22,572 cubic inches
- ○ D. 21,289 cubic inches

29. The park has two L-shaped community swimming pools, pool A and pool B. Each pool has a child section and an adult section. The volume of the child section of pool A is 144 cubic feet. The volume of the adult section for pool A is 12,660 cubic feet. What is the total volume, in cubic feet, of pool A?

Part A:

- ○ A. 12,804 cubic feet
- ○ B. 68 cubic feet
- ○ C. 13,188 cubic feet
- ○ D. 6,594 cubic feet

Part B: Pool B has the same volume as pool A. The volume of the child section of pool B is 253 cubic feet. What is the volume of the adult section, in cubic feet, for pool B? Write your answer in the box below.

cubic feet

30. Write your answer in the box.

$5 \times (6 + 2) \div 4 =$ ☐

31. Write the expression for *25 less than the product of 53 and 5* in the box below.

☐

32. Select the correct written statement for the expression below.

$$7 \times (52 + 256)$$

- ○ A. Seven added to fifty-two plus two hundred fifty-six
- ○ B. Seven times the sum of fifty-two and two hundred fifty-six
- ○ C. Seven times fifty-two minus two hundred fifty-six
- ○ D. Seven more than fifty-two minus two hundred fifty-six

33. Which **two** statements about the corresponding terms in both Pattern A and Pattern B is always true?

Pattern A: 0, 1, 2, 3, 4, 5

Pattern B: 0, 4, 8, 12, 16, 20

- □ A. Each term in Pattern A is 2 less than the corresponding term in Pattern B.
- □ B. Each term in Pattern A is $\frac{1}{4}$ of the corresponding term in Pattern B.
- □ C. Each term in Pattern B is 4 times the corresponding term in Pattern A.
- □ D. Each term in Pattern A is 4 times the corresponding term in Pattern B.

34. Graph points *A*, *B*, and *C* on the coordinate plane. Point *A* should be located at (5, 7), point *B* should be located at (3, 6), and point *C* should be located at (2, 7). Graph all three points.

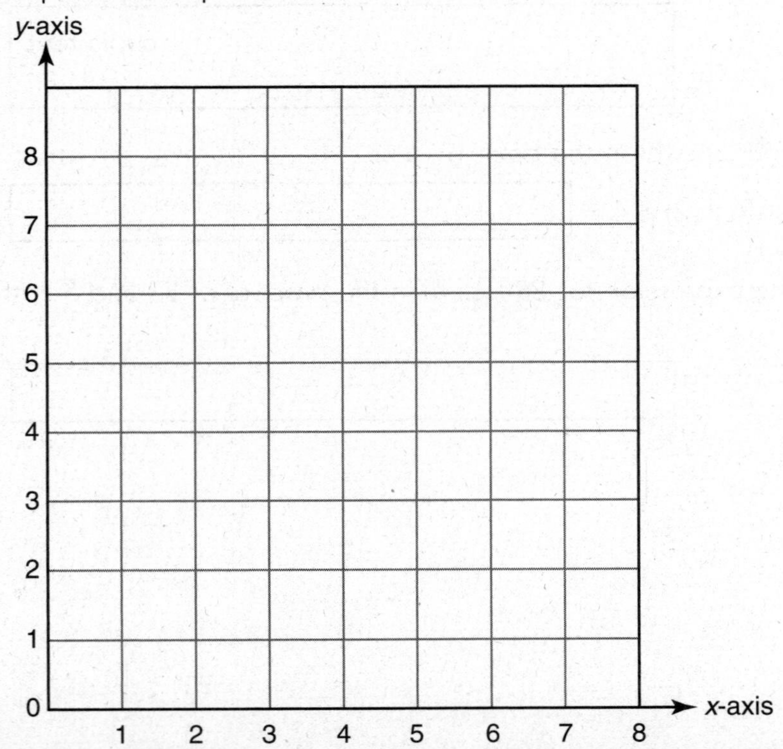

35. Which explanation about figures is **not** correct?

- O A. A square is a trapezoid because it has exactly one pair of parallel sides.
- O B. A square is a parallelogram because it has two pairs of parallel sides.
- O C. A square is a rectangle because it has four right angles, and the opposite sides are equal and parallel.
- O D. A square is a rhombus because it has four sides and two pairs of parallel sides.

36. Use your knowledge of classifying quadrilaterals to answer the questions below:

	YES	NO
A. All squares are rectangles.		
B. All quadrilaterals are trapezoids.		
C. All parallelograms are rectangles.		
D. All squares are rhombuses.		
E. All parallelograms are quadrilaterals.		
F. All rectangles are rhombuses.		

PRACTICE TEST 2—ANSWERS EXPLAINED

Question	Answer	CCSS	Detailed Explanation
1	B	5.NBT.2	10 is multiplied by itself 5 times. The exponent 5 tells how many times to use 10 as a factor, known as a power of 10. Note that the power of 10 has the same number of zeros as the exponent.
2	B, D	5.NBT.3a	701.269 is written in word-form and expanded form.
3	<	5.NBT.3b	In comparing the numbers the values are the same until we reach the hundredths place. The value of $9 \times \frac{1}{1,000}$ in the number 56.839 is greater since we do not have a digit in the thousandths place of the number 56.83.
4	10,360	5.NBT.5	
5	891,149	5.NBT.5	
6	.57 57	5.NBT.int.1	Task does not require you to multiply but to understand place value system.
7	255.05	5.NBT.7-1	
8	5.456	5.NBT.7-3	
9	13	5.NBT.7-4	Remember $\frac{10}{10} = 1$ whole. You have 52 tenths of some object that needs to be divided into 4 tenth sections. How many groups of 4 tenths do you have? Skip counting by 4 you arrive at the answer of 13. 4 tenths, 8 tenths, 12 tenths, 16 tenths, 20 tenths, 24, 28, 32, 36, 40, 42, 48, 52 tenths

Question	Answer	CCSS	Detailed Explanation
10	$\frac{4}{12}$ or $\frac{1}{3}$	5.NF.1	
11	$\frac{31}{24}$ or $1\frac{7}{24}$	5.NF.1-2	Find common denominators or use equivalent fractions.
12	$\frac{7}{20}$	5.NF.1-3	Find common denominators or use equivalent fractions.
13	A, C	5.NF.1-5	Find common denominators or use equivalent fractions.
14	B	5.NF.2-2	Recognize that $\frac{5}{6}$ is almost 1.
15	Part A: $1\frac{19}{24}$ lb Part B: $4\frac{7}{24}$ lb	5.NF.2-1	Use your knowledge of finding a common denominator to subtract the fractions. Use your knowledge of finding the common denominator to add and subtract fractions.
16	D	5.NF.3-1	$\frac{2}{5} = 2 \div 5$ Two hot dogs divided equally among five friends.
17	Part A: $23\frac{1}{3}$ Part B: 23 and 24	5.NF.3-2	350 boxes divided equally among 15 people.
18	A	5.NF.4a-1	$\frac{a}{b} \times q = \frac{aq}{b}$ Remember the formula to multiply a fraction times a whole number: $\frac{a}{b} \times q = \frac{a \times q}{b} = a \times q \div b$ You jumped rope for $\frac{1}{4}$ of an hour for 5 days. This can be written as $\frac{1}{4} \times 5 = \frac{1}{4} + \frac{1}{4} + \frac{1}{4} + \frac{1}{4} + \frac{1}{4} = \frac{5}{4}$ hours or $\frac{1}{4} \times 5 = 1 \times 5 \div 4 = \frac{5}{4}$ hours

Question	Answer	CCSS	Detailed Explanation
19	$\frac{14}{24}$ or $\frac{7}{12}$	5.NF.4a-2	$\frac{a}{b}\times\frac{c}{d}=\frac{ac}{bd}$ Remember the formula to multiply a fraction times a fraction: $\frac{a}{b}\times\frac{c}{d}=\frac{a\times c}{b\times d}=\frac{ac}{bd}$ What part of the class had cotton candy? Seven-eighths of the class had cotton candy and two-thirds of the cotton candy eaters got sick. Using your formula you can arrive at a solution. $\frac{7}{8}\times\frac{2}{3}=\frac{14}{24}=\frac{7}{12}$
20	A. Greater than B. Less than C. Equal to	5.NF.5a	
21	C	5.NF.6-1	$\frac{1}{8}\times\frac{2}{3}=\frac{2}{24}=\frac{1}{12}$
22	\$294	5.NF.6-2	One approach to solving the problem $32\frac{2}{3}\times\$9=?$ $\frac{98}{3}\times\$9=\frac{98\times9}{3}=\frac{882}{3}=\$294$

Question	Answer	CCSS	Detailed Explanation
23	63	5.NF.7c	This problem is similar to question 18. You have to remember how to multiply and divide by fractions. One approach: You have 21 yards of fabric that need to be divided into $\frac{1}{3}$ yard pieces. How many groups of $\frac{1}{3}$ do you have? Note: $\frac{3}{3} = 1$ whole or 1 yard. Each yard will make 3 pennants. 21 yards × 3 = 63 pennants. 1 whole = $\frac{3}{3}$
24	A. 0.5 m B. 5,000 m C. 50,000 cm	5.MD.1	Remember to convert to a smaller unit, move the decimal point to the right or multiply by a power of 10. (see table below) To convert to a larger unit, move the decimal point to the left and divide by a power of 10.

	km	hm	dk	m	dm	cm	mm
A				0.5	5	50	
B	5	50	500	5000			
C	0.5	5	50	500	5000	50,000	

Question	Answer	CCSS	Detailed Explanation
25	Part A 1,760 ounces or 13.75 gallons Part B 88 Students	5.MD.1-2	1 gallon = 128 ounces 1 quart = 2 pints 1 pint = 16 ounces PART A 15 gallons × 128 ounces = 1,920 ounces 5 quarts = 10 pints = 160 ounces 1920 ounces − 160 ounces = 1760 ounces remaining. Students may choose to answer as 13.75 gallons or represent it as a mixed fraction. PART B 1.25 pints = 20 ounces 1760 ounces ÷ 20 ounces = 88 students
26	Part A: See Detailed Explanation. Part B: $\frac{31}{96}$ in	5.MD.2-2	0, $\frac{1}{8}$, $\frac{1}{4}$, $\frac{1}{2}$, $\frac{3}{4}$, 1 Use knowledge of adding and dividing fractions to arrive at your answer. $\left(\frac{3}{2}+\frac{5}{4}+\frac{3}{4}+\frac{3}{8}\right)\div 12=$
27	24 cubic centimeters	5.MD.4	Count the blocks. 24 blocks means volume is 24 cubic centimeters.
28	C	5.MD.5b	
29	Part A: A Part B: 12,551 cubic feet	5.MD.5c	
30	10	5.OA.1	
31	$(53 \times 5) - 25$	5.OA.2-1	

<table>
<tr><th>Question</th><th>Answer</th><th>CCSS</th><th>Detailed Explanation</th></tr>
<tr><td>32</td><td>B</td><td>5.OA.2-2</td><td></td></tr>
<tr><td>33</td><td>B, C</td><td>5.OA.3</td><td>Pattern A is $\frac{1}{4}$ the values in Pattern B, and Pattern B represents 4 times the values in Pattern A.</td></tr>
<tr><td>34</td><td>See Detailed Explanation.</td><td>5.G.1</td><td> 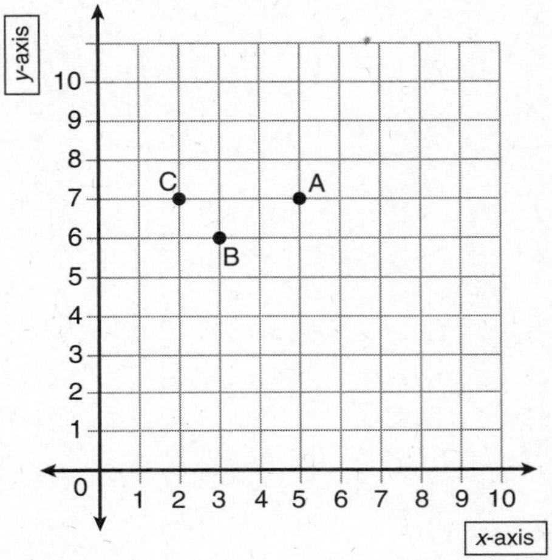</td></tr>
<tr><td>35</td><td>A</td><td>5.G.3</td><td>A square has exactly two pairs of parallel sides and, thus, is not a trapezoid.</td></tr>
<tr><td>36</td><td>See Detailed Explanation.</td><td>5.G.4</td><td><table>
<tr><th></th><th>YES</th><th>NO</th></tr>
<tr><td>All squares are rectangles.</td><td>X</td><td></td></tr>
<tr><td>All quadrilaterals are trapezoids.</td><td></td><td>X</td></tr>
<tr><td>All parallelograms are rectangles.</td><td></td><td>X</td></tr>
<tr><td>All squares are rhombuses.</td><td>X</td><td></td></tr>
<tr><td>All parallelograms are quadrilaterals.</td><td>X</td><td></td></tr>
<tr><td>All rectangles are rhombuses.</td><td></td><td>X</td></tr>
</table></td></tr>
</table>

Web-Based Resources

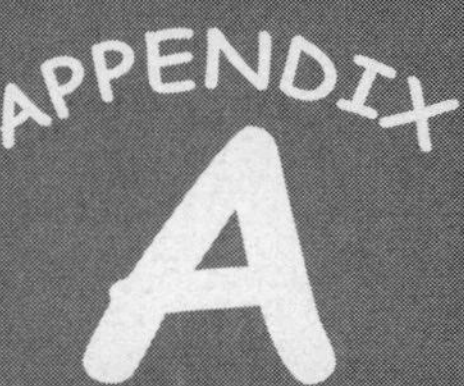

As of the publication date of this book the following websites were available:

http://commoncoresheets.com

http://splashmath.com- Limited access of 20 questions per day

http://www.arcademicskillbuilders.com/

http://www.corestandards.org/standards-in-your-state/

https://www.khanacademy.org

http://www.k-5mathteachingresources.com

http://www.learn-with-math-games.com/

http://www.mathchimp.com/

http://www.mathfactcafe.com/grade/5/

http://www.sheppardsoftware.com/math.htm

http://www.sumdog.com/

Base-Ten Block Sheets

*Note: Make copies of this page and use for extra practice.

*Note: Make copies of this page and use for extra practice.

Place Value Charts

	Hundreds	Tens	Ones	•	Tenths	Hundredths	Thousandths
				•			
				•			

	Hundreds	Tens	Ones	•	Tenths	Hundredths	Thousandths
				•			
				•			

	Hundreds	Tens	Ones	•	Tenths	Hundredths	Thousandths
				•			
				•			

	Hundreds	Tens	Ones	•	Tenths	Hundredths	Thousandths
				•			
				•			

	Hundreds	Tens	Ones	•	Tenths	Hundredths	Thousandths
				•			
				•			

	Hundreds	Tens	Ones	•	Tenths	Hundredths	Thousandths
				•			
				•			

*Note: Make copies of this page and use for extra practice.

Thousands	Hundreds	Tens	Ones	Tenths	Hundredths	Thousandths

Thousands	Hundreds	Tens	Ones	Tenths	Hundredths	Thousandths

*Note: Make copies of this page and use for extra practice.

Index

O

P

Q

R

S

T

U

V

W

X

Y

Put New Jersey Students on the Path to Success

NEW JERSEY STATE ASSESSMENT TESTS

Preparing students for success is so important. That's why these newly revised study guides are the perfect tools to help students get ready for the PARCC (The Partnership for Assessment of Readiness for College and Careers) tests in the State of New Jersey. Through a series of high quality, computer-based K–12 assessments in Mathematics and English Language Arts, teachers, schools, and parents can rest assured that students are on track for success. Each title features:

- Two full-length practice tests with answers and explanations
- In-depth review through engaging lessons, hints, and tips for each PARCC test
- Content reflective of classroom lessons and aligned to Common Core Curriculum
- An explanation and overview of the PARCC assessment

Give students the help they need to achieve their very best—now and in the future.

BARRON'S
NEW JERSEY
GRADE 5
ELA/LITERACY TEST
Mark Riccardi, M.Ed.
Kimberly Perillo, M.Ed.
Reflective of the NEW PARCC ASSESSMENT
2 full-length practice tests with answers and explanations
Fully aligned to the Common Core Curriculum
Covers all fifth-grade English Language Arts topics prescribed by the State of New Jersey
Provides an explanation and overview of the PARCC assessments

New Jersey Grade 5 ELA/Literacy Test
Mark Riccardi and Kimberly Perillo
ISBN 978-1-4380-0560-7, $12.99, *Can$14.99*

New Jersey Grade 5 Math Test
Stephenie Tidwell
ISBN 978-1-4380-0722-9, $14.99, *Can$17.99*

Each book: Paperback, 7 13/16" x 10"

Available at your local book store
or visit **www.barronseduc.com**

Barron's Educational Series, Inc.
250 Wireless Blvd.
Hauppauge, NY 11788
Order toll-free: 1-800-645-3476

In Canada: Georgetown Book Warehouse
34 Armstrong Ave.
Georgetown, Ont. L7G 4R9
Canadian orders: 1-800-247-7160

Prices subject to change without notice.

(#156b) R9/15

Your Key to **COMMON CORE SUCCESS**

The recent implementation of Common Core Standards across the nation has offered new challenges to teachers, parents, and students. The ***Common Core Success*** series gives educators, parents, and children a clear-cut way to meet—and exceed—those grade-level goals.

Our English Language Arts (ELA) and Math workbooks are specifically designed to mirror the way teachers actually teach in the classroom. Each workbook is arranged to engage students and reinforce the standards in a meaningful way. This includes:

- Units divided into thematic lessons and designed for self-guided study
- "Stop and Think" sections throughout the ELA units, consisting of "Review," "Understand," and "Discover"
- "Ace It Time!" activities that offer a math-rich problem for each lesson

Students will find a wealth of practical information to help them master the Common Core!

Barron's Common Core Success
Grade K English Language Arts/Math
978-1-4380-0668-0

Barron's Common Core Success
Grade 1 English Language Arts
978-1-4380-0669-7

Barron's Common Core Success
Grade 1 Math
978-1-4380-0670-3

Barron's Common Core Success
Grade 2 English Language Arts
978-1-4380-0671-0

Barron's Common Core Success
Grade 2 Math
978-1-4380-0672-7

Barron's Common Core Success
Grade 3 English Language Arts
978-1-4380-0673-4

Barron's Common Core Success
Grade 3 Math
978-1-4380-0674-1

Barron's Common Core Success
Grade 4 English Language Arts
978-1-4380-0675-8

Barron's Common Core Success
Grade 4 Math
978-1-4380-0676-5

Barron's Common Core Success
Grade 5 English Language Arts
978-1-4380-0677-2

Barron's Common Core Success
Grade 5 Math
978-1-4380-0678-9

Barron's Common Core Success
Grade 6 English Language Arts
978-1-4380-0679-6

Barron's Common Core Success
Grade 6 Math
978-1-4380-0680-2

COMMON CORE SUCCESS WORKBOOKS GRADES K–6

Each book:
Paperback,
8 3/8" x 10 7/8"
$12.99, *Can$15.50*

Available at your local book store
or visit **www.barronseduc.com**

Barron's Educational Series, Inc.
250 Wireless Blvd.
Hauppauge, N.Y. 11788
Order toll-free: 1-800-645-3476

Prices subject to change without notice.

In Canada:
Georgetown Book Warehouse
34 Armstrong Ave.
Georgetown, Ontario L7G 4R9
Canadian orders:
1-800-247-7160

(#293) R3/15

GRADES 2–7 TEST PRACTICE for Common Core

With Common Core Standards being implemented across America, it's important to give students, teachers, and parents the tools they need to achieve success. That's why Barron's has created the *Core Focus* series. These multi-faceted, grade-specific workbooks are designed for self-study learning, and the units in each book are divided into thematic lessons that include:

- Specific, focused practice through a variety of exercises, including multiple-choice, short answer, and extended response questions
- A unique scaffolded layout that organizes questions in a way that challenges students to apply the standards in multiple formats
- "Fast Fact" boxes and a cumulative assessment in Mathematics and English Language Arts (ELA) to help students increase knowledge and demonstrate understanding across the standards

Perfect for in-school or at-home study, these engaging and versatile workbooks will help students meet and exceed the expectations of the Common Core.

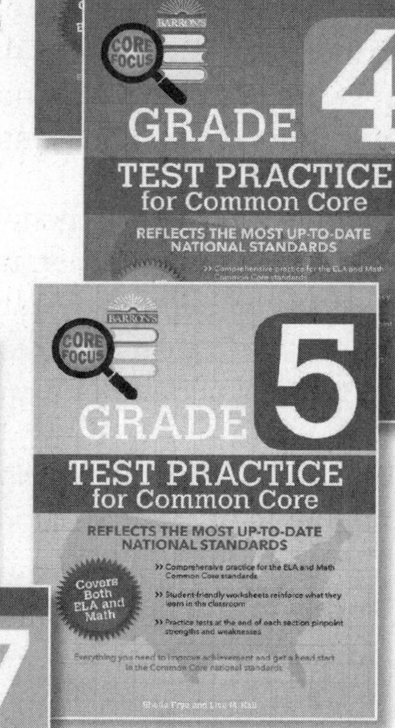

Grade 2 Test Practice for Common Core
Maryrose Walsh and Judith Brendel
ISBN 978-1-4380-0550-8
Paperback, $14.99, *Can$16.99*

Grade 3 Test Practice for Common Core
Renee Snyder, M.A. and Susan M. Signet, M.A.
ISBN 978-1-4380-0551-5
Paperback, $14.99, *Can$16.99*

Grade 4 Test Practice for Common Core
Kelli Dolan and Shephali Chokshi-Fox
ISBN 978-1-4380-0515-7
Paperback, $14.99, *Can$16.99*

Grade 5 Test Practice for Common Core
Lisa M. Hall and Sheila Frye
ISBN 978-1-4380-0595-9
Paperback, $14.99, *Can$16.99*

Grade 6 Test Practice for Common Core
Christine R. Gray and Carrie Meyers-Herron
ISBN 978-1-4380-0592-8
Paperback, $14.99, *Can$16.99*

Grade 7 Test Practice for Common Core
Techla Connolly and Carrie Meyers-Herron
ISBN 978-1-4380-0706-9
Paperback, $14.99, *Can$17.99*

Coming soon to your local book store or visit **www.barronseduc.com**

Barron's Educational Series, Inc.
250 Wireless Blvd.
Hauppauge, N.Y. 11788
Order toll-free: 1-800-645-3476

In Canada:
Georgetown Book Warehouse
34 Armstrong Ave.
Georgetown, Ontario L7G 4R9
Canadian orders: 1-800-247-7160

Prices subject to change without notice.

(#295 R8/15)

How kids can earn, save, and invest their own money . . .

MONEY $ENSE for Kids!

Second Edition

by Hollis Page Harman, P.F.P.

Teaching young children about money is a skill-set they will value for the rest of their lives. Boys and girls can find answers to dozens of intriguing questions about money, starting with . . .

- How and where is money printed?
- What do all those long numbers and special letters on currency mean?
- How can banks afford to pay interest?
- Can kids find savings programs designed especially for them?

They'll also have fun solving puzzles and playing investment games that focus on the theme of money. Diagrams and illustrations on most pages. (Ages 8 and older)

Paperback • ISBN 978-0-7641-2894-3 • $14.99, *Can$17.99*

To order —— Available at your local book store
or visit **www.barronseduc.com**

Barron's Educational Series, Inc.
250 Wireless Blvd.
Hauppauge, N.Y. 11788
Order toll-free: 1-800-645-3476
Order by fax: 1-631-434-3217

In Canada:
Georgetown Book Warehouse
34 Armstrong Ave.
Georgetown, Ontario L7G 4R9
Canadian orders: 1-800-247-7160
Order by fax: 1-800-887-1594

Prices subject to change without notice.

(#256) R4/12